cooking from
the market
vegetables

cooking from the market
vegetables

MURDOCH BOOKS

contents

introduction

In a growing trend towards buying food that is fresh and in season, more
of us are shopping locally from grower's markets. We want to reconnect
with our food and the people who grow it. *Cooking from the Market* gives
you tips on what to look for when buying your vegetables and how to
store and prepare them. And, most importantly, it features a selection
of recipes that showcase each vegetable to perfection.

winter vegetables

beetroot

Beetroot (beet) is notoriously messy to prepare, leaving dark red stains on anything its juice comes into contact with, and it also takes quite a long time to cook — perhaps two of the reasons why so many cooks opt for the convenient tinned version. If you've only ever tasted beetroot from tins, give the fresh version a try — it might open your eyes to a whole new world.

Buying and storing

- Beetroot is available year-round but its peak season is from winter to spring.

- Always buy beetroot with the stems and leaves attached. The leaves should look fresh and bright and the bulbs should be firm and deep purple in colour.

- If possible, choose bulbs that are roughly the same size or they will all have different cooking times. Avoid any that are very big, as these may be woody.

- Store trimmed beetroot loosely tied in a plastic bag in the crisper of the fridge for 7–10 days. Use the leaves and stems within 2 days of purchase.

Preparation

First cut off the leaves, leaving 3 cm (1¼ inches) of stalk, and leaving the roots attached. If boiling the beetroot, scrub the beetroot gently to remove the dirt, then cook them in their skins, as this preserves their flavour and colour (if you peel them first, much of the colour bleeds into the cooking water). For deep-frying or roasting, recipes may suggest to peel the beetroot first.

Boiling Cook large to medium beetroot in simmering salted water for up to 1½ hours, or until tender. Pierce them with a skewer to check they are cooked. Small beetroot will take 20–30 minutes to cook.

When the cooked beetroot are cool enough to handle, peel off the skins — wear rubber gloves to prevent purple-stained hands.

Once cooked, beetroot can be sliced or cut into wedges and used in salads, risottos, dips and vegetable side dishes. Beetroot can also be grated raw and added to salads, or thin slices of raw beetroot can be deep-fried to make 'chips'.

Beetroot hummus

SERVES 8

250 g (9 oz/1 cup) dried chickpeas
1 large onion, chopped
500 g (1 lb 2 oz) beetroot (beets)
140 g (5 oz/^{1}/$_2$ cup) tahini (sesame
  seed paste)
3 garlic cloves, crushed
3 tablespoons lemon juice
1 tablespoon ground cumin
3 tablespoons olive oil

Put the chickpeas in a large bowl, cover with cold water and soak overnight. Drain the chickpeas and put them in a large heavy-based saucepan with the onion. Cover with water and bring to the boil. Cook for 1 hour, or until the chickpeas are very soft. Drain, reserving 250 ml (9 fl oz/1 cup) of cooking liquid. Allow the chickpeas to cool.

Cut the leaves off the beetroot bulbs, leaving 3 cm (1^1/$_4$ inches) of stalk attached, but don't trim the roots. Wash well to remove all traces of soil. Cook the beetroot in a large saucepan of boiling salted water for 1–1^1/$_2$ hours, or until tender when tested with a skewer. Drain and cool slightly before removing the skins.

Chop the beetroot and place in a food processor, in batches if necessary. Add the chickpea and onion mixture, tahini, garlic, lemon juice and cumin, and process until smooth. Slowly add the reserved cooking liquid and the olive oil while the machine is running. Blend until the mixture is thoroughly combined. Drizzle with a little olive oil and serve with crudités or country–style bread.

Beetroot and cumin salad

SERVES 4–6

6 beetroot (beets)
4 tablespoons olive oil
1 tablespoon red wine vinegar
^{1}/$_2$ teaspoon ground cumin
1 red onion
1 large handful flat-leaf (Italian) parsley,
  chopped

Cut the leaves off the beetroot bulbs, leaving 3 cm (1^1/$_4$ inches) of stalk attached, but don't trim the roots. Wash well to remove all traces of soil. Cook the beetroot in boiling salted water for 1–1^1/$_2$ hours, or until tender when tested with a skewer. Drain and set aside to cool slightly.

In a bowl, beat the olive oil with the red wine vinegar, cumin and some freshly ground black pepper to make a dressing.

Wearing rubber gloves so the beetroot juice doesn't stain your hands, peel the warm beetroot bulbs and trim the roots. Halve the bulbs and cut into slender wedges and place in the dressing in the bowl. Halve the onion, slice into slender wedges and add to the beetroot. Add the parsley and toss well. Transfer to a serving plate and serve warm or at room temperature.

Beetroot chips

SERVES 4 AS A SNACK

500 g (1 lb 2 oz) beetroot (beets),
    peeled
750 ml (26 fl oz/3 cups) vegetable oil
1/2 quantity aïoli (page 67), to serve

Use a sharp knife to cut the beetroot into
paper-thin slices. Preheat the oven to 120°C
(235°F/Gas 1/2).

Heat the oil in a deep frying pan over high
heat and cook the beetroot slices, in batches,
until they are crisp and browned. Drain on
paper towels and keep warm in the oven
while cooking the remainder.

Serve the beetroot chips with the aïoli,
for dipping.

Roasted beetroot with horseradish cream

SERVES 4–6 AS A SIDE DISH

8 beetroot (beets)
2 tablespoons olive oil
2 teaspoons honey
chopped parsley, to garnish

horseradish cream
150 ml (5 fl oz) thick (double/heavy) cream
2 teaspoons grated fresh horseradish
2 teaspoons lemon juice
a pinch of caster (superfine) sugar

Preheat the oven to 200°C (400°F/Gas 6).
Scrub and peel the beetroot, trim the ends
and cut into quarters. Put the oil and honey
in a bowl and mix well. Season, then divide
the beetroot pieces among four large squares
of foil. Drizzle with the honey mixture, turning
to coat well, then wrap loosely in the foil.
Bake for 1 hour, or until the beetroot are
tender when pierced with a skewer.

To make the horseradish cream, whip the
cream until it just starts to thicken, then fold
in the horseradish, lemon juice, sugar and
a pinch of salt.

Once the beetroot are cooked, remove from
the oven and leave in the foil for 5 minutes.
Serve with a generous dollop of horseradish
cream, and garnish with parsley.

Small **beetroot leaves** can be used in salads or as you would
spinach, or try them chopped and stewed in olive oil with garlic,
anchovies, currants and a splash of red wine **vinegar.**

carrot

Carrots are an indispensable ingredient in kitchens the world over; so versatile as to be used in dishes as varied as cakes and casseroles, in stocks and stews, raw in salads and as a delicious juice. While it's hard to imagine carrots being any colour other than orange, the original carrots were white, yellow, pink, black or purple. The orange version didn't appear until the 1700s and was developed by the Dutch, believed to have been motivated out of patriotism for the ruling House of Orange.

Buying and storing

- Although carrots are available year-round, their peak season is from autumn through to spring.

- Buy carrots with smooth, unblemished skin, with no wrinkled spots or splits in the skin. The deeper the orange colour, the higher the beta-carotene content.

- Avoid carrots with green 'shoulders', as these have been exposed to sunlight and will be bitter. Don't buy carrots that are too big either, as these may have a woody core.

- If you buy a bunch of carrots with their green tops still attached, cut these off for storage, as the tops will drain the carrots of nutrients and moisture.

- Store in a plastic bag in the crisper of the fridge for up to 1 week.

Preparation

Unlike many other root vegetables, carrots are delicious both raw and cooked. Baby carrots, sometimes called Dutch carrots, are those that have been either bred to mature when small, or are a larger variety that have been thinned out of the main crop. These are sold in bunches with their green, feathery tops still on. These carrots don't need peeling, just scrub them, then trim off the tops, leaving a small green stem for presentation, and serve them raw or steam whole. Older, larger carrots are best peeled and cooked.

Roasting While we tend to usually just boil carrots, they are very good when roasted, as this concentrates and caramelises their sugars. Slice into rounds, then put in a roasting tin with a drizzle of olive oil. Roast for 30 minutes in a 180°C (350°F/Gas 4) oven, turning once, until deep golden. Serve tossed in balsamic vinegar and sprinkled with sea salt and chopped mint.

Carrots contain more vitamin A than any other vegetable (they contain beta-carotene, which converts to vitamin A in the body). This vitamin, among other things, is beneficial for function of the retina.

Carrot cake with lemon icing

SERVES 8

125 g (4$^{1}/_{2}$ oz/1 cup) self-raising flour
125 g (4$^{1}/_{2}$ oz/1 cup) plain (all-purpose) flour
2 teaspoons ground cinnamon
1 teaspoon ground ginger
$^{1}/_{2}$ teaspoon ground nutmeg
1 teaspoon bicarbonate of soda (baking soda)
250 ml (9 fl oz/1 cup) oil
185 g (6$^{1}/_{2}$ oz/1 cup) soft brown sugar
4 eggs
175 g (6 oz/$^{1}/_{2}$ cup) golden syrup or dark corn syrup
400 g (14 oz/2$^{1}/_{2}$ cups) grated carrot
60 g (2$^{1}/_{4}$ oz/$^{1}/_{2}$ cup) chopped pecans or walnuts

lemon icing
175 g (6 oz/$^{3}/_{4}$ cup) cream cheese, softened
60 g (2$^{1}/_{4}$ oz) unsalted butter, softened
185 g (6$^{1}/_{2}$ oz/1$^{1}/_{2}$ cups) icing (confectioners') sugar
1 teaspoon natural vanilla extract
1–2 teaspoons lemon juice
ground nutmeg, to dust

Preheat the oven to 160°C (315°F/Gas 2–3). Grease a 23 cm (9 inch) round cake tin and line the base and side with baking paper.

Sift the flours, cinnamon, ginger, nutmeg and bicarbonate of soda into a large bowl and make a well in the centre. In a small bowl, whisk together the oil, sugar, eggs and golden syrup until combined. Add this mixture to the well in the flour and gradually stir into the dry ingredients with a metal spoon until smooth. Stir in the carrot and nuts, mixing thoroughly.

Spoon the batter into the prepared tin and smooth the surface. Bake for 1$^{1}/_{2}$ hours, or until a skewer inserted into the centre of the cake comes out clean. Leave the cake in the tin for at least 15 minutes before turning out onto a wire rack to cool completely.

To make the lemon icing, beat the cream cheese and butter with electric beaters until smooth. Gradually add the icing sugar alternately with the vanilla and lemon juice, beating until light and creamy. Spread the icing over the cooled cake using a flat-bladed knife. Sprinkle with the ground nutmeg.

Spicy carrot soup

SERVES 4

2 tablespoons olive oil
1 onion, chopped
800 g (1 lb 12 oz) carrots, chopped
1 bay leaf
1 teaspoon ground cumin
1 teaspoon cayenne pepper
1 teaspoon ground coriander
2 teaspoons paprika
1.25 litres (44 fl oz/5 cups) vegetable stock
250 g (9 oz/1 cup) Greek-style yoghurt
1 small handful coriander (cilantro)

Heat the olive oil in a saucepan, add the onion and carrot and cook over low heat for 30 minutes.

Add the bay leaf, cumin, cayenne pepper, coriander and paprika and cook for another 2 minutes. Add the stock, bring to the boil, then reduce the heat and simmer, uncovered, for 40 minutes, or until the carrot is tender. Cool slightly, then blend in batches in a food processor. Return the soup to the saucepan and gently reheat. Season with sea salt and freshly ground black pepper.

Combine the yoghurt with the coriander leaves. Ladle the soup into bowls and top with a dollop of the yoghurt mixture.

Orange and carrot salad

SERVES 6

3 sweet oranges
500 g (1 lb 2 oz) carrots, julienned
2 tablespoons lemon juice
1 teaspoon ground cinnamon, plus extra
    to serve
1 tablespoon caster (superfine) sugar
1 tablespoon orange flower water
small mint leaves, to serve

Cut off the tops and bases of the oranges. Cut the skin off, removing all the white pith and cutting through the outer membranes to expose the flesh. Holding the orange over a bowl to catch the juice, segment the oranges by cutting between the membranes. Discard the seeds and place the segments in the bowl, cover and refrigerate. Squeeze the membranes to extract all the juice, then pour the juice into another bowl.

Put the carrots in the bowl with the orange juice. Add the lemon juice, cinnamon, sugar, orange flower water and a small pinch of salt. Stir well to combine. Cover and refrigerate.

Just before serving, drain off the juice from the oranges and arrange the segments around the edge of a serving plate. Pile the carrots in the centre and top with the mint. Dust the oranges with a little extra cinnamon.

Carrots go well with apples, oranges, honey, butter, brown sugar, chives, cumin, mint, parsley, cream, walnuts and raisins.

Carrot and pumpkin risotto

SERVES 4

90 g (3 1/4 oz) butter
1 onion, finely chopped
250 g (9 oz) pumpkin (winter squash), diced
2 carrots, diced
2 litres (70 fl oz/8 cups) vegetable stock
440 g (15 1/2 oz/2 cups) risotto (arborio) rice
90 g (3 1/4 oz/1 cup) shaved romano or
    parmesan cheese
1/4 teaspoon ground nutmeg

Heat 60 g (2 1/4 oz) of the butter in a large heavy-based frying pan. Add the onion and fry for 1–2 minutes over medium heat, or until the onion is soft, then add the pumpkin and carrot and cook for 6–8 minutes, or until tender. Mash slightly with a potato masher. In a separate saucepan, heat the stock over medium heat and keep the stock at simmering point.

Add the rice to the vegetables and cook for 1 minute, stirring constantly, until the rice grains are translucent. Ladle in 125 ml (4 fl oz/1/2 cup) of the hot stock and stir well. Reduce the heat and when the rice has absorbed all of the stock, add the next ladleful of stock, stirring constantly, for 20–25 minutes, or until the rice is tender and creamy. You may not need to add all the stock, or you may run out and need to use a little more stock or water — every risotto is different.

Remove from the heat, add the remaining butter, the cheese and nutmeg. Season with freshly ground black pepper and stir well to combine. Cover and leave for 5 minutes before serving.

Carrot and almond salad

SERVES 4

4 large carrots
2 tablespoons peanut oil
1 teaspoon caster (superfine) sugar
1/2 teaspoon brown mustard seeds
1/4 teaspoon curry powder
2 tablespoons lemon juice
25 g (1 oz/1/4 cup) flaked almonds, toasted
1 large handful coriander (cilantro)
60 g (2 1/4 oz/1/4 cup) Greek-style yoghurt

Preheat the grill (broiler) to medium. Slice the carrots thinly on the diagonal. Put half the peanut oil in a bowl and add the sugar, stirring to dissolve, then add the carrots and toss to coat.

Spread the carrots on a baking tray and grill (broil) for 10–15 minutes, turning occasionally, until lightly browned and tender. Remove the tray from the heat and leave the carrots to cool, then transfer the carrots to a bowl.

While the carrots are cooking, heat the remaining oil in a small frying pan. Add the mustard seeds and curry powder and cook over low heat for 1 minute, or until fragrant. Allow to cool a little, then whisk in the lemon juice and season with salt and freshly ground black pepper. Drizzle the spice mixture over the carrots, add the almonds and coriander and toss gently until well combined. Serve at room temperature, with a dollop of yoghurt.

celeriac

Related to celery and carrots, celeriac is a bumpy, gnarled-looking vegetable with a tangle of roots at its base. Contrary to its appearance, celeriac has a delicate celery-like flavour and is thoroughly delicious. Use it raw, cut into matchsticks and tossed with mayonnaise in the classic French salad, celeriac rémoulade; in soups and stews; or mashed with garlic and potatoes. Good partners for its mildly aniseed taste include watercress, crumbled goat's cheese, walnuts, hazelnuts, apples, oranges, capers, anchovies, olives and boiled eggs.

Buying and storing

- Celeriac is in season during autumn and winter, and is sometimes available in spring.

- Avoid buying celeriac that are overly large, as these tend to have cottony centres. Select bulbs that are very heavy for their size, feel firm and are about the size of a baseball, as these will have the crispest, most dense interior. For easier peeling, choose smoother roots. If they still have their leaves attached, these should be bright green.

- Trim the leaves off the bulbs and store in a bag in the crisper of the fridge for 7–10 days.

Preparation

Preparing celeriac can seem like a wasteful exercise, as quite a lot of trimming and cutting is needed. First cut off and discard the stalks and leaves, then remove all of the thick skin — you need to get rid of all the tangled roots at the base. Put cut pieces of celeriac into a bowl of water with a squeeze of lemon juice, as the flesh quickly oxidises on contact with the air.

Boiling Peeled chunks of celeriac can be boiled in salted water for about 15 minutes, or until tender. They can then be mashed with butter and cream, or tossed with a little butter and black pepper and served as a side dish with sausages, smoked meats, chicken or fish.

Deep-frying Celeriac can also be cut into matchsticks and deep-fried in hot oil until golden and crisp (parboil it first for quicker cooking), and served with fish, chicken or beef.

Roasting Peel and cut the celeriac into 1.5 cm (⅝ inch) wedges. Toss generously in olive oil or butter, then arrange in a large roasting tin in a single layer. Cook in a 200°C (400°F/Gas 6) oven for 20–25 minutes, or until tender and golden.

Celeriac and potato mash

SERVES 4 AS A SIDE DISH

1 tablespoon lemon juice
1 celeriac
1 large roasting potato, such as russet (idaho)
    or king edward, peeled and cut into
    2.5 cm (1 inch) chunks
250 ml (9 fl oz/1 cup) milk
20 g (3/4 oz) butter, softened

Pour 500 ml (17 fl oz/2 cups) cold water into a large bowl and add the lemon juice. Trim and peel the celeriac, then chop into 2.5 cm (1 inch) chunks, placing them in the acidulated water to prevent discolouration.

Drain well, then place in a saucepan with the potato and milk and bring to the boil over high heat. Cover and cook for 15 minutes, or until the celeriac and potato are tender. Mash well, season to taste with sea salt and freshly ground black pepper, then stir in the butter and serve.

variation To make celeriac and apple mash, add 2 peeled, cored and coarsely chopped granny smith apples to the potato and celeriac, then boil with the milk and continue as above.

Celeriac rémoulade

SERVES 4–6 AS A STARTER OR SIDE DISH

mustard mayonnaise
2 egg yolks
1 tablespoon white wine vinegar
    or lemon juice
1 tablespoon dijon mustard
150 ml (5 fl oz) olive oil

juice of 1 lemon
1 large or 2 small celeriac
2 tablespoons capers, rinsed and drained
5 gherkins (pickles), chopped
2 tablespoons finely chopped parsley
crusty bread, to serve

To make the mustard mayonnaise, whisk the egg yolks, vinegar and mustard together in a bowl. Whisking constantly, add the olive oil, 1 teaspoon at a time, until the mixture begins to thicken, then add the remaining oil in a thin, steady stream. Set aside until required, placing plastic wrap directly on the surface of the mayonnaise to prevent a skin forming.

Pour 1 litre (35 fl oz/4 cups) cold water into a large bowl and add half the lemon juice. Trim and peel the celeriac, then coarsely grate and place in the acidulated water to prevent discolouration. Bring a saucepan of water to the boil over high heat and add the remaining lemon juice. Drain the celeriac and add to the saucepan. After 1 minute, drain the celeriac again and cool under running water. Pat dry with paper towels.

Toss the celeriac with the mayonnaise, capers, gherkins and parsley. Serve with crusty bread.

eggplant

Eggplants (aubergines) are used in cooking as a vegetable but they are actually a fruit and a member of the same family as tomatoes and potatoes. Because of the number of eggplant dishes that hail from the Mediterranean — there's eggplant parmigiano and caponata from Italy, ratatouille from France, moussaka from Greece, and Turkey is reputed to have over 1000 eggplant recipes — it is often thought that this was where they originated. So it may come as a surprise to learn that the eggplant is native to Asia, and probably came from India or Burma.

Varieties

There are many eggplant varieties, varying in size and shape, with colours ranging from green, cream or yellow to pink-and-white striped, pale or dark purple. Some of the most commonly grown varieties are listed below.

Western or globe The classic, large, dark purple eggplant, these vary in size from around 350 g to 550 g (12 oz to 1 lb 4 oz). They have few seeds but as they age, their flesh becomes darker, the seeds become more numerous and the flavour deteriorates into bitterness. Slice and use for frying or grilling (broiling), or cut into chunks for roasting, stewing or sautéing. Many cooks prefer to peel them before use as the skin can be tough.

Japanese These long, slim eggplants are usually purple-skinned or sometimes light mauve. They have a mildly sweet flavour and because their skin is thin, there's no need to peel them. Use them for stuffing and baking, pickling, chargrilling, stir-frying, grilling (broiling) and steaming.

Pea These small eggplants grow in clusters and they may be coloured red, orange, green or purple. They are quite bitter, and are often used in pickles and curries and other Asian dishes.

Thai These small, round eggplants are lavender in colour with green striations, or green or cream. They have tough skin, a seedy interior and a bitter flavour. Use unpeeled and sliced in Thai soups, salads and curries.

Buying and storing

- Eggplants are available year-round from markets, making them a versatile vegetable for both summer salads and side dishes, as well as wintery dishes such as gratins, moussakas and tagines.

- Eggplants are quite perishable, so it is important to choose them at their peak. They should be firm and heavy for their size and have taut, shiny, smooth skins with no browning or soft spots. When you press the skin of a larger eggplant, a dent should form but should disappear soon after.

- Don't store eggplants in the fridge but at cool room temperature for 2–3 days. Eggplants can brown and alter in flavour if refrigerated.

Preparation

Degorging Most recipes suggest to salt eggplants before cooking them, and this is known as degorging. The reason generally given for this is that the salt draws out any bitterness, but today most commercially grown eggplants are not bitter unless they are old. However, salting can also reduce the amount of oil the eggplants absorb as they cook (blanching in boiling water can also help with this).

To salt, remove the green stem, then slice the eggplant into the desired thickness or cut into chunks. Layer these in a colander, scattering with salt as you go, then stand the colander over a sink for 30–40 minutes to drain. Rinse the eggplant well in cold water, drain and pat dry on paper towels. Use as directed in your recipe.

Baking Toss chunks or slices of eggplant in oil, arrange in a baking dish and cook in a 180°C (350°F/Gas 4) oven for 40–50 minutes, until very soft and golden. Note that eggplant is one vegetable that should never be undercooked — it needs to be cooked to the point that the flesh is almost falling apart. You can also bake a whole eggplant in this way. Prick the eggplant several times with a skewer, then brush lightly with oil. Cook for 45 minutes, or until very soft. Allow to cool before peeling.

Roasting First use a metal skewer to prick a few holes over the unpeeled eggplant, then cook it directly over a gas flame, turning often, for 25 minutes, or until the skin is blackened and the eggplant is soft. When cool enough to handle, remove all of the skin, then drain the flesh well in a colander, squeezing to remove excess moisture. This results in wonderfully smoky-flavoured eggplant (use in dips and spreads) but can be messy, so put foil around the gas element to catch juices.

Grilling Cut the eggplant into slices and brush with oil. Cook over medium heat on a barbecue grill or chargrill pan for about 3 minutes each side, or until tender. Alternatively, brush with oil and place under the grill (broiler) and cook for 3–5 minutes on each side.

Eggplant parmigiana

SERVES 6–8 AS A SIDE DISH OR LIGHT MEAL

1.25 kg (2 lb 12 oz) tomatoes
olive oil
1 onion, diced
2 garlic cloves, crushed
1 kg (2 lb 4 oz) eggplants (aubergines),
    thinly sliced
250 g (9 oz) bocconcini (fresh baby
    mozzarella cheese), sliced
185 g (6^1/$_2$ oz/1^1/$_2$ cups) finely grated
    cheddar cheese
1 large handful basil, torn
50 g (1^3/$_4$ oz/^{1}/$_2$ cup) grated parmesan
    cheese

Bring a saucepan of water to the boil. Using a small knife, score a cross in the base of each tomato. Place the tomatoes in the boiling water for about 20 seconds, remove using a slotted spoon, then plunge into a bowl of iced water. Drain the tomatoes and peel the skins away from the cross. Cut the tomatoes in half, scoop out the seeds with a teaspoon and roughly chop the flesh.

Heat 3 tablespoons of olive oil in a large saucepan. Add the onion and sauté over medium heat for 5 minutes, or until softened. Add the garlic and cook for 1 minute, then add the tomato and simmer for 15 minutes. Season to taste with sea salt.

Meanwhile, preheat the oven to 200°C (400°F/Gas 6).

Place a large frying pan over medium heat and pour in enough olive oil to cover the base. Add the eggplant and fry in batches for 3–4 minutes, or until golden brown, adding more oil to the pan as needed. Drain on paper towels.

Place one-third of the eggplant in a 1.75 litre (61 fl oz/7 cup) baking dish. Top with half the bocconcini and cheddar. Repeat the layers, finishing with a layer of eggplant. Pour the tomato mixture over the top, then scatter with the basil and parmesan. Transfer to the oven and bake for 40 minutes, or until the eggplant is tender. Serve hot.

variation If you would prefer not to fry the eggplant in oil, brush it with olive oil and brown lightly under a hot grill (broiler).

Moroccan eggplant with couscous

SERVES 4 AS A SIDE DISH

185 g (6½ oz/1 cup) instant couscous
200 ml (7 fl oz) olive oil
1 onion, halved and sliced
1 eggplant (aubergine)
1 teaspoon garlic salt
3 teaspoons ground cumin
¼ teaspoon ground cinnamon
1 teaspoon paprika
¼ teaspoon ground cloves
50 g (1¾ oz) butter
1 handful flat-leaf (Italian) parsley,
    roughly chopped

Put the couscous in a large bowl and add 375 ml (13 fl oz/1½ cups) boiling water. Cover and leave for 10 minutes, then fluff up the grains with a fork.

Heat 2 tablespoons of the olive oil in a large frying pan. Add the onion and sauté over medium heat for 8–10 minutes, or until browned. Remove from the pan, leaving the pan on the stovetop to use again.

Cut the eggplant into 1 cm (½ inch) slices, then cut the slices into quarters. Place in a large bowl. In a small bowl, mix together the garlic salt, cumin, cinnamon, paprika, cloves and ½ teaspoon sea salt, then sprinkle over the eggplant, tossing to coat well.

Add the remaining oil to the pan and reheat the pan over medium heat. Add the eggplant and cook, turning once, for 20 minutes, or until soft and browned. Remove from the pan and set aside to cool.

Melt the butter in the frying pan. Add the couscous and cook, stirring, for 2–3 minutes, then stir in the onion, eggplant and parsley. Serve at room temperature with roast lamb or chicken.

The **eggplant** is a member of the deadly nightshade family and was once viewed with great suspicion. The Spanish called it the 'apple of love' as they believed it was an aphrodisiac. The Italians also, cautiously, developed an appreciation for this new vegetable, but in northern Europe and England it was used ornamentally only — eating it was thought to induce insanity (perhaps this is why some called it the 'mad apple'). The English gave it the name we use today, as the first eggplants they saw were small, egg-shaped and creamy white in colour.

fennel

Fennel looks like a bulbous bunch of celery; it has the same crunchy texture but has an aniseed flavour. Use thinly sliced raw fennel in salads, teamed with ingredients such as capers, oranges, watercress and parmesan. Lightly cooked and tossed through pasta, it goes well with cream and lemon, and braised fennel is great with a cheesy sauce.

Preparation

To serve raw, trim the tops and base of the fennel with a small knife, discarding any tough, fibrous outer leaves. Cut the bulb in half lengthways and trim the core, then slice thinly for salads, or into wider strips for crudités. Prepare fennel just before using, or place in a bowl of water with a squeeze of lemon juice to prevent it turning brown.

For braising and roasting, cut halved, trimmed bulbs into wedges, but don't trim the core as this will hold the fennel together. Fennel can also be chopped or sliced, then sautéed for use in sauces, vegetable dishes or stews.

Roasting Brush wedges of fennel with olive oil, put in a lightly greased baking dish and season with salt and pepper. Cook in a 180°C (350°F/ Gas 4) oven for 35–40 minutes, turning once, or until tender.

Braising Place some wedges of fennel in a baking dish and pour over some hot chicken stock to nearly cover. Add some olive oil and season with fresh thyme, salt and pepper. Cover with foil and cook in a 180°C (350°F/Gas 4) oven for 35–45 minutes, or until tender.

Buying and storing

- Although bulb fennel is available year-round, it is best from mid-autumn to winter.
- Never buy fennel without its tops as these indicate freshness. The tops should be bright green, and the bulbs should be tight, firm and greenish white in colour. Don't buy any bulbs that have flabby, loose outside layers.
- Small bulbs of fennel will have a milder flavour than larger ones, so choose these if you want to eat them raw.
- Store fennel in a plastic bag in the fridge for up to 4 days.

Florentine roast pork

SERVES 6

Pork loin roasted with garlic and rosemary is served all over Tuscany, traditionally roasted on a spit over a fire. In this version, the pork is served on a bed of soft fennel and onion.

3 large fennel bulbs
2 teaspoons finely chopped rosemary
4 garlic cloves, crushed
1.5 kg (3 lb 5 oz) pork loin, chined and
  skinned
3 white onions
90 ml (3 fl oz) olive oil
185 ml (6 fl oz/¾ cup) dry white wine
4 tablespoons extra virgin olive oil
250 ml (9 fl oz/1 cup) chicken stock
3–4 tablespoons thick (double/heavy) cream

Preheat the oven to 200°C (400°F/Gas 6). Cut the green fronds from the tops of the fennel and chop to give 2 tablespoons fronds. Add these to a small bowl along with the rosemary, garlic and plenty of salt and freshly ground black pepper.

Make deep incisions with a sharp knife all over the pork and rub this mixture into the incisions and the splits in the pork bone. Cut two of the onions in half and place in a roasting tin. Put the pork on top of the onion and drizzle the olive oil over the top.

Roast in the oven for 30 minutes. Baste the pork with the pan juices, then reduce the oven to 180°C (350°F/Gas 4) and roast for a further 30 minutes. Baste and lightly salt the surface of the pork, then pour in half the wine. Continue roasting for another 30–45 minutes, basting once or twice.

Meanwhile, remove the tough outer layers of the fennel and discard. Slice the bulbs vertically into 1 cm (½ inch) sections and place in a large saucepan. Thinly slice the remaining onion and add to the saucepan along with the extra virgin olive oil and a little salt. Add enough water to cover. Put a lid on the saucepan and bring to the boil. Simmer for about 45 minutes, or until the fennel is creamy and soft and almost all the liquid has evaporated.

Remove the pork from the tin and leave to rest. Spoon off the excess oil from the tin and discard the onion. Place the tin over high heat on the stovetop and stir in the remaining wine to deglaze. Add the stock and boil the sauce until slightly thickened.

Remove from the heat, season with salt and freshly ground black pepper and stir in the cream. Slice the pork and serve on the fennel. Serve with the sauce.

In 1824, an American official living in the Italian city of Florence, sent some **fennel seeds** to the former U.S. president, Thomas Jefferson, along with a gushing letter: 'Fennel is beyond every other vegetable, delicious …' The Americans, along with various other nations, were slow to catch on to what the Italians have known for a long time — that fennel is a wonderful and versatile vegetable.

Fennel, tomato and garlic gratin

SERVES 6

900 g (2 lb) fennel bulbs (about 2 bulbs)
4 tablespoons olive oil
1 large red onion, halved and thinly sliced
2 garlic cloves, crushed
450 g (1 lb) tomatoes

gratin topping
60 g (2 1/4 oz/3/4 cup) fresh breadcrumbs
65 g (2 1/2 oz/2/3 cup) grated parmesan
    cheese
2 teaspoons finely grated lemon zest
1 garlic clove, crushed

Preheat the oven to 200°C (400°F/Gas 6). Grease a 22 cm (8 1/2 inch) square gratin dish. Cut the fennel in half lengthways, then slice thinly.

Heat the olive oil in a large frying pan. Cook the onion for 3–4 minutes, or until softened but not browned, then add the garlic and cook for 2 minutes. Add the fennel and cook, stirring frequently, for 7 minutes, or until softened and golden brown.

Using a small knife, score a cross in the base of each tomato. Place the tomatoes in boiling water for about 20 seconds, remove using a slotted spoon, then plunge into a bowl of iced water. Drain the tomatoes and peel the skins away from the cross. Chop the tomato flesh roughly and add to the fennel. Cook, stirring frequently, for 5 minutes, or until the tomato has softened. Season well and tip into the gratin dish.

To make the gratin topping, mix together the breadcrumbs, parmesan, lemon zest and garlic. Sprinkle over the vegetables and bake for 15 minutes, or until golden brown and crisp. Serve immediately.

Spaghetti with sardines, fennel and tomato

SERVES 4–6

3 roma (plum) tomatoes
4 tablespoons olive oil
80 g (2¾ oz/1 cup) fresh breadcrumbs
3 garlic cloves, crushed
1 red onion, thinly sliced
1 fennel bulb, quartered and thinly sliced
3 tablespoons raisins
3 tablespoons pine nuts, toasted
4 anchovy fillets, chopped
125 ml (4 fl oz/½ cup) white wine
1 tablespoon tomato paste (concentrated purée)
4 tablespoons finely chopped flat-leaf (Italian) parsley
350 g (12 oz) butterflied sardines (ask your fishmonger to do this)
500 g (1 lb 2 oz) spaghetti

Bring a saucepan of water to the boil. Using a small knife, score a cross in the base of each tomato. Place the tomatoes in the boiling water for about 20 seconds, remove using a slotted spoon, then plunge into a bowl of iced water. Drain the tomatoes and peel the skins away from the cross. Cut the tomatoes in half and scoop out the seeds with a teaspoon, then roughly chop the flesh.

Heat 1 tablespoon of the olive oil in a large frying pan over medium heat. Add the breadcrumbs and one-third of the garlic and cook, stirring often, for 5 minutes, or until the breadcrumbs are golden and crisp. Remove from the pan using a slotted spoon.

Heat the remaining oil in the same pan. Add the onion, fennel and remaining garlic and sauté for 8 minutes, or until the vegetables are soft. Add the chopped tomato, raisins, pine nuts and anchovies and cook for a further 3 minutes, then stir in the wine, tomato paste and 125 ml (4 fl oz/½ cup) water. Simmer for 10 minutes, or until the mixture thickens slightly. Stir in the parsley, then set aside and keep warm.

Pat the sardines dry with paper towels. Cook the sardines in batches in a lightly greased frying pan over medium heat for 1–2 minutes, or until just cooked through. Take care not to overcook or they will break up. Set aside.

Meanwhile, cook the spaghetti in a large saucepan of rapidly boiling salted water until al dente. Drain and return to the pan.

Add the sauce to the spaghetti and stir to coat well, then add the sardines and half the fried breadcrumbs and toss to combine. Sprinkle the remaining breadcrumbs over the top and serve immediately.

mushroom

Mushrooms are not technically vegetables — they are the fleshy, fruiting body of a form of fungus, which grow in soil or decaying matter. There are countless varieties of mushrooms, some cultivated, others gathered from the wild — ranging in size, shape, flavour and texture — but only some of these are edible.

Edible fungi can be broadly divided into three categories: cultivated mushrooms, wild mushrooms and exotic mushrooms.

Cultivated mushrooms

These mushrooms are available year-round and are grown in compost in temperature- and moisture-controlled environments. All modern-day cultivated mushrooms belong to the *Agaricus bisporus* family, the only difference being that they are harvested at different stages of maturity.

Button The most commonly available mushroom, these have a mild flavour that intensifies when cooked, but they are also delicious raw in salads. Sauté whole then marinate in olive oil, use for cream of mushroom soup or in casseroles, sauces and pies.

Chestnut These look similar to button mushrooms but are darker in colour, with open caps. They have a meaty, firm texture and strong, nutty flavour, and are suited for eating raw, or lightly cook and use in pastry fillings, pâtés and breads.

Portobello Growing up to 10 cm (4 inches) in diameter, these are the largest of the cultivated mushrooms. They have a meaty texture and rich flavour and are ideal for barbecuing or pan-frying whole, or make a stuffing of breadcrumbs, sage, pancetta and grated parmesan, fill the caps with the mixture and then bake them.

Swiss brown Known as criminis in the United States, swiss browns are actually the original form of the common button mushroom, not a variety in their own right. They have a more intense flavour than button mushrooms and are good in meat or game dishes and pasta sauces, or cook chopped mushrooms in oil and garlic and use as a crostini topping or frittata filling.

Wild mushrooms

Wild mushrooms rely upon the cooler weather to flourish and are generally in season during autumn and winter.

Chanterelles These frilly golden-hued mushrooms are prized for their subtle fruity flavour and firm texture. Use in pasta dishes, risottos, sauces for meat, with chicken and pheasant, or sauté in butter and serve with scrambled eggs. Add towards the end of cooking to prevent them from toughening.

Field These open and flat-capped mushrooms grow in fields and cow pastures from late winter to early spring. Enjoy them pan-fried, crumbed and deep-fried, roasted, grilled (broiled), stuffed, barbecued or added to braised meat dishes. If used in braised poultry recipes, they can turn the dish an unappetising grey, although this does not affect their flavour.

Morel Closely related to the truffle, morels are one of the few spring mushrooms. Relatively scarce, they are highly prized for their earthy, smoky, nut-like flavour. The darker the morel, the more intensely flavoured it is. They add a depth of flavour to game, veal and chicken dishes and are commonly cooked with cream. Morels are also available dried.

Pine These mushrooms grow wild at the roots of pine trees. This category includes slippery jacks, which have a delicate flavour, moist slippery caps and spongy flesh, and saffron milk caps, so-named because of their saffron tones and the milky liquid they emit. They have a nutty flavour and firm texture and are excellent when combined with other mushrooms and used in sauces, soups and stews.

Porcini (ceps) Indigenous to France and Italy's alpine regions, porcinis are loved for their distinctly sweet, strong mushroom flavour. Available fresh (only during late summer and autumn), dried or frozen, their pungent aroma invigorates any dish they are added to. Use in risottos, pasta sauces, soups and stews.

Exotic mushrooms

Once exclusive to Japan, China and Korea, where they originated, these 'exotic' mushrooms are now increasingly available fresh in most greengrocers and some supermarkets.

Enoki These mushrooms grow in clusters that are linked at the base, and have tiny caps and slender long stems. With a sweet flavour and crunchy texture, these delicate mushrooms require only minimal cooking, so add at the last minute to Japanese soups and hotpots, or use raw in salads.

Oyster These pale beige to grey mushrooms have fluted fan-shaped caps. Cooking brings out their peppery, delicate flavour — fry in butter and garlic, use in Japanese-style braises, add to casseroles and clear soups, or brush the mushrooms with tamari and sake and barbecue as an accompaniment for teriyaki beef.

Shiitake Indigenous to Japan, shiitakes are now the most widely cultivated of all the Asian mushrooms and are available fresh and dried. Shiitakes have broad, plump and firm caps. Use in braises, sautés, stir-fries, roasts and soups, or finely chop and use in fillings for dumplings.

Shimeji Confusingly, 'shimeji' refers to about 20 mushroom species. The hon-shimeji is regarded as the 'true' shimeji and was first cultivated in Japan. It has a nutty, slightly sweet flavour and is good in soups, stir-fries and sauces, but is not suitable for eating raw. Shimeji are often sold in a dense cluster (this is how they grow); cut off the base and wipe them before use — do not wet them or they will become soggy.

It is crucial to always forage for **mushrooms** with an experienced mushroom hunter or ask a professional to identify any suspect mushrooms before eating them, as there are many more poisonous varieties than edible ones and mistakes can be fatal.

Buying and storing

- Cultivated and exotic mushrooms are available from the market year-round; wild mushrooms are sold from autumn to winter.

- Choose mushrooms with no signs of tearing or soft spots. The stems should still be intact and the caps firm and not shrivelled or dry, or 'tacky', which indicates spoilage.

- If you are planning to cook the mushrooms whole, choose ones that are similar in size so they cook in the same amount of time.

- Store mushrooms in a paper bag in the fridge for up to 3 days. The paper bag prevents them from dehydrating.

- Mushrooms absorb other flavours, so store them away from strong-smelling foods.

- Exotic mushrooms are generally sold on plastic-wrapped trays or in disposable containers. Choose mushrooms that look plump and firm, then transfer to a perforated plastic or paper bag and refrigerate on a shelf, not the crisper, for up to 3 days.

Preparation

Mushrooms are porous and readily absorb water, so don't wash them before storing or cooking. Trim the stems (or remove them completely if you are going to stuff the mushroom), then wipe with a damp cloth to remove any dirt. Morels, however, need to be thoroughly washed as bugs can hide in their hollow centres.

Roasting A good method for cooking large-capped mushrooms is to place them in a baking dish, skin side down, drizzle with oil and scatter over some herbs such as thyme or marjoram. Season well, then cook in a 200°C (400°F/Gas 6) oven for 12–15 minutes, turning once.

Sautéing First slice or chop the mushrooms, then cook in oil in a large frying pan over medium–high heat for 5–7 minutes, or until soft and golden. Make sure the oil is hot before adding the mushrooms and be aware that they will absorb the oil, so don't be tempted to add more as they will release juices and the soaked-up oil as they soften.

Grilling To grill (broil) or barbecue mushrooms, brush with oil and place under a medium–high grill (broiler) for 7–10 minutes, turning once.

Mushroom pâté on mini toasts

MAKES 24

60 g (2¼ oz) butter
1 small onion, chopped
3 garlic cloves, crushed
375 g (13 oz) button mushrooms, quartered
125 g (4½ oz/1 cup) slivered almonds, toasted
2 tablespoons cream
2 tablespoons finely chopped thyme
3 tablespoons finely chopped flat-leaf (Italian) parsley
6 thick slices wholemeal (whole-wheat) bread

Heat the butter in a large frying pan. Add the onion and garlic and cook over medium heat for 2 minutes, or until soft. Increase the heat, add the mushrooms and cook for 5 minutes, or until the mushrooms are soft and most of the liquid has evaporated. Leave to cool for 10 minutes.

Roughly chop the toasted almonds in a food processor. Add the mushroom mixture and process until smooth. With the motor running, gradually pour in the cream. Stir in the thyme and parsley and season with sea salt and freshly ground black pepper. Spoon into two 250 ml (9 fl oz/1 cup) ramekins and smooth the surface. Cover and refrigerate for 4–5 hours to let the flavours develop.

To make the toasts, preheat the oven to 180°C (350°F/Gas 4). Toast one side of the bread under a hot grill (broiler) until golden. Remove the crusts and cut each slice into four triangles. Place on a large baking tray in a single layer, toasted side down, and cook for 5–10 minutes, or until crisp. Serve immediately with the pâté.

Baked mushrooms

SERVES 4 AS A SIDE DISH

250 g (9 oz) button mushrooms
200 g (7 oz) oyster mushrooms
200 g (7 oz) fresh shiitake mushrooms
100 g (3½ oz) swiss brown mushrooms
2 tablespoons extra virgin olive oil

topping
80 g (2¾ oz/1 cup) fresh breadcrumbs (see note)
3 tablespoons grated parmesan cheese
2 tablespoons chopped flat-leaf (Italian) parsley
1 tablespoon chopped thyme
2 garlic cloves, crushed

Preheat the oven to 180°C (350°F/Gas 4). Trim the mushroom stems, then wipe off any dirt using a damp cloth or paper towel. Cut any large mushrooms in half.

Sprinkle the base of a large baking dish with a little water. Place the mushrooms in a single layer in the dish, stems facing upwards.

Put all the topping ingredients in a bowl along with 1 teaspoon freshly ground black pepper and mix together. Sprinkle the topping over the mushrooms and drizzle with the olive oil. Bake for 12–15 minutes, until the mushrooms have heated through. Serve warm.

note Slightly stale day-old white bread is perfect for making breadcrumbs. Remove the crusts and chop the bread in a food processor until crumbs form. Avoid using pre-sliced packaged bread for making breadcrumbs, as this type of bread tends not to go stale. For best results, use baguette, sourdough or a similar bread from a bakery.

The Egyptian pharaohs regarded **mushrooms** as an exclusive royal food and declared that commoners were not allowed to eat them. Other ancients considered that mushrooms contained all manner of magical properties — some believed that eating them gave them superhuman powers, while others thought mushrooms powerful enough to help them recover lost objects.

Open lasagne of mushrooms, pine nuts and thyme

SERVES 4

200 g (7 oz) fresh pasta sheets
   (2 medium sheets)
80 g (2¾ oz) butter
1 tablespoon olive oil
300 g (10½ oz) assorted mushrooms,
   sliced
2 bacon slices, cut into pieces of similar
   size to the mushroom slices
2 garlic cloves, thinly sliced
1 tablespoon thyme
1 tablespoon pine nuts, toasted
3 tablespoons thick (double/heavy) cream
3 tablespoons extra virgin olive oil
35 g (1¼ oz/⅓ cup) coarsely grated
   pecorino cheese

Bring a large saucepan of salted water to the boil. Cut the pasta sheets into sixteen 8 cm (3¼ inch) squares. Boil half of the squares for 4 minutes, or until al dente. Using a slotted spoon, transfer the squares to a bowl of cold water, leave for 15–20 seconds, then drain. Lay the pasta flat on a dry tea towel and cover with another tea towel; it doesn't matter that the squares may have cooked to uneven sizes. Repeat this process with the remaining pasta squares.

Heat the butter and olive oil in a large frying pan. Add the mushrooms and bacon to the pan and cook over high heat, stirring often, for 3–4 minutes, or until golden brown. Add the garlic and thyme and cook for a further 1 minute. Add the pine nuts, cream and 2 tablespoons of extra virgin olive oil and stir until combined. Remove from the heat and season to taste.

Preheat the grill (broiler) to medium–high. Put a pasta square in each of four shallow heatproof pasta bowls. Cover with a heaped tablespoon of mushroom mixture. Repeat twice more, then top each one with the fourth pasta square; the pasta doesn't have to be in uniform stacks, nor the piles neat.

Drizzle the remaining tablespoon of extra virgin olive oil over the top and scatter with the cheese. Place the bowls under the heat for 1–2 minutes, or until the cheese has melted. Serve hot or warm.

parsnip

Before the potato came along, the parsnip was Europe's most popular winter vegetable. Parsnips need richness to bring out their silky texture and nutty flavour, so roast them in butter or olive oil, bake them or purée them with cream. They also pair beautifully with sausages, bacon, and roasted or grilled lamb or beef.

Buying and storing

- Parsnips are at their best during the colder months, as their distinctive flavour relies on the icy snap of winter; after the first frosts they become sweeter, as the cold causes them to convert their starch into sugar.

- Parsnips are always sold without their green tops, as these contain an irritating juice that can cause a serious rash on some people.

- Buy parsnips with smooth, cream-coloured skin and avoid any with browning on their skin or with tops that have started to sprout.

- Parsnips come in a variety of shapes — some are very elongated and tapering, while others are fat and stumpy; choose the fatter ones as there will be less wastage. Select those that are firm and feel heavy for their size and try to pick them of a similar size so they will all cook at the same time.

- Store parsnips in the fridge for up to 1 week. Don't wrap them in plastic, as they will sweat and deteriorate.

Preparation

Cut larger parsnips in half lengthways and inspect the core; very large or old parsnips may need their woody cores removed. Peel them just before cooking, as they discolour slightly on contact with the air.

Parsnips often need to be precooked before using them in a mash or baking into a gratin, for example. Boiling them isn't always the best cooking option as they can become water-logged; instead, try steaming them, or tossing them in olive oil and roasting in a 200°C (400°F/ Gas 6) oven for 30–45 minutes, or until golden and tender.

Parsnip and leek purée

SERVES 4–6 AS A SIDE DISH

1 leek, white part only, thinly sliced
3 large parsnips, peeled and chopped
2 tablespoons snipped chives
30 g (1 oz) butter
3 tablespoons crème fraîche

Put the leek and parsnips in a saucepan of boiling salted water. Cook for 10 minutes, or until tender. Drain well and allow to cool slightly before blending in a food processor to form a smooth purée.

Put the purée in a clean saucepan and add the chives and butter. Season with salt and freshly ground black pepper. Cook over low heat until the purée is heated through, then remove from the heat and stir through the crème fraîche. Serve with grilled lamb cutlets, roast beef or pork.

Parsnip and mustard soup

SERVES 4–6

30 g (1 oz) butter
1 onion, chopped
750 g (1 lb 10 oz) parsnips, chopped
1 litre (35 fl oz/4 cups) vegetable stock
125 ml (4 fl oz/$1/2$ cup) milk
125 ml (4 fl oz/$1/2$ cup) cream
2–3 tablespoons wholegrain mustard
2 tablespoons chopped flat-leaf (Italian) parsley

Melt the butter in a large saucepan, add the onion and cook over medium heat, stirring often, until soft but not brown.

Add the parsnips and stock and bring to the boil. Simmer, covered, for 25 minutes, or until tender. Set aside to cool slightly.

Blend the soup in batches in a blender or food processor. Return to the pan, add the milk and cream and reheat gently, but do not allow the soup to boil. Stir in the mustard and season to taste with salt and freshly ground black pepper. Garnish with parsley to serve.

Parsnips have been used as a food for thousands of years. They were so esteemed that the Roman emperor Tiberius had them brought to him all the way from France and Germany. The Romans adored them, and even ate them for dessert with honey and fruits. In the New World, the early settlers used parsnips in pies and stews, to make chips, in breads and puddings, and even used them to make wine.

potato

From Chile to China, from Ireland to Idaho, the humble potato has given countless societies endless ways to enjoy its comforting goodness. Potatoes are a staple in the global diet and after rice, wheat and corn they are the world's fourth largest food crop. Dependable and adaptable, the potato is equally tempting whether mashed with cream, roasted with rosemary until golden brown, deep-fried in oil to make crispy chips or baked to tenderness in its jacket and topped with sour cream and chives.

Varieties

There are hundreds of varieties of potatoes, and these are called different things in different countries, so for cooking purposes, we have broadly grouped them according to their 'type'.

Floury These dry potatoes have a low sugar content but are high in starch and are perfect for baking, mashing and frying. They include russet (idaho), spunta, coliban, pentland squire and king edward.

Waxy These potatoes have moist, dense flesh that is low in starch. They are ideal for boiling, and in salads or stews, as they won't lose their shape when cooked. Most new potatoes, red-skinned potatoes and many of the kipfler (fingerling) varieties fall into this group. Kipflers are small, long potatoes. They have a firm texture, making them ideal for steaming and boiling and for use in salads. Other waxy potatoes include roseval, charlotte, pink fir apple and cara.

All-purpose Many of the varieties of potatoes sold in supermarkets fall into this category. These have been specially bred to span all possible uses. These include desiree, sebago, nicola, kennebecs, maris piper, bintje (yellow fin), spunta, pontiac and pink eye.

New and old potatoes

Newly dug potatoes, sold as 'new' potatoes, are harvested from plants that still have their green leaves, usually during winter. The skin of these slightly immature potatoes is fragile and thin, and their flesh is dense and moist. They need only to be boiled or steamed, with little more than sea salt or butter as flavourings.

Maincrop potatoes, or 'old' potatoes, are those harvested when their vines are blooming. These have drier flesh and thicker skins than new potatoes. They are usually stored in cool, dark conditions to extend their availability (most supermarket potatoes are 'old'). Any potato variety, however, can be harvested new or old.

Buying and storing

- Although potatoes are available year-round, they are best during autumn and winter.

- Avoid potatoes that feel soft or spongy or have any discolouration or bruises. Don't buy any that are starting to sprout, as these have been stored for too long.

- Store potatoes in a dark, dry place; if stored correctly, they will keep for 10–12 weeks. Avoid refrigeration, plastic bags and sunlight, as these will promote softening, sprouting and spoiling.

- New potatoes and small thin-skinned varieties should be used within a few days.

Preparation

Cut off any green areas on your potatoes (this may only be visible after you've peeled them) as these contain a toxic substance called solanine. The greening occurs through excessive exposure to light; similarly, cut off any eyes that are starting to sprout. Unless cooking them immediately, put peeled or cut potatoes into a bowl of water, as the flesh will discolour slightly.

Much of the potato's nutritional goodness (mainly its protein content) is stored just under the skin, so don't peel them unless you need them for a potato salad or mash, for example. Alternatively, boil them in their skins, then peel when cool enough to handle.

Boiling Place potatoes, either whole and unpeeled or peeled and cut, into a saucepan and cover with cold water. Add salt and bring to the boil, then reduce the heat and simmer, uncovered, for 15–20 minutes, or until tender. Drain well or they will become soggy.

Baking Scrub unpeeled potatoes, prick them several times with the point of a knife, then place directly on a rack in a 200°C (400°F/Gas 6) oven. Bake for 50 minutes to 1¼ hours, depending on their size, until tender.

Roasting Parboil small unpeeled potatoes or halved medium-sized potatoes for 20 minutes, or until tender but still firm. Drain well and leave in the colander for 15 minutes to allow the moisture to evaporate. Toss potatoes into the roasting tin with whatever meat you are roasting, turning them occasionally so they brown and crisp in the cooking fats and juices — this will take about 50 minutes in a 180°C (350°F/Gas 4) oven. To roast them on their own, toss in olive oil and cook as above.

To make **potato wedges**, scrub and pat dry 8 floury potatoes and cut into thick wedges. Put in a single layer on a baking tray, brush lightly with olive oil and sprinkle with sea salt and sweet paprika (or chilli powder). Cook in a 220°C (425°F/Gas 7) oven for 35–40 minutes, or until crisp and golden. Serve with sweet chilli sauce or aïoli (page 67) as a dipping sauce.

Potato croquettes

MAKES 12

750 g (1 lb 10 oz) floury potatoes, coarsely
    chopped
2 tablespoons cream or melted butter
3 eggs, lightly beaten
1/4 teaspoon ground nutmeg
plain (all-purpose) flour, for coating
150 g (5 1/2 oz/1 1/2 cups) dry breadcrumbs
vegetable oil, for deep-frying

Put the potatoes in a saucepan and cover
with water. Add a little salt, then bring to the
boil and cook for 10 minutes, or until tender.
Drain and return the potato to the hot pan to
dry out a little, then mash with a fork. Stir in
the cream, a third of the beaten egg and the
nutmeg. Season with sea salt and freshly
ground black pepper. Spread the potato
mixture evenly on a baking tray, cover and
place in the fridge for 30 minutes.

Divide the mixture into 12 equal portions and
roll each portion into a sausage shape. Roll
in the flour, shaking off the excess. Dip each
croquette in the remaining egg, then coat
in the breadcrumbs, shaking off any excess.
Cover and refrigerate for at least 2 hours.

Half-fill a deep heavy-based frying pan
with oil and heat to 180°C (350°F), or until
a cube of bread dropped into the oil browns
in 15 seconds. Cook the croquettes in batches
for 5 minutes each, or until golden. Drain on
paper towels and serve.

Potato cake

SERVES 4 AS A SIDE DISH

1 small onion, sliced into rings
75 g (2 1/2 oz) butter
2 garlic cloves, crushed
1 kg (2 lb 4 oz) potatoes, thinly sliced
100 g (3 1/2 oz/2/3 cup) grated mozzarella
    cheese
50 g (1 3/4 oz/1/2 cup) shaved parmesan
    cheese
2 tablespoons milk

Put the onion in a bowl, cover with cold water
and leave for 1 hour. Drain well. Preheat the
oven to 210°C (415°F/Gas 6–7). Line a 20 cm
(8 inch) spring-form cake tin with foil. Grease
the foil.

Melt the butter in a small saucepan, add the
garlic and set aside. Put a layer of potato over
the base of the tin, followed by layers of
onion, butter, mozzarella and parmesan
cheeses. Repeat the layers until you have
used up all the ingredients, finishing with
potato and keeping a bit of butter to drizzle
over at the end. Season the layers as you go,
then spoon the milk over the top.

Bake for 1 hour, or until the top is golden
brown and the potatoes are tender. If the top
is browning too much before the potatoes are
cooked, cover with foil. Cool for 10 minutes
before serving. Unclip the base of the tin,
peel off the foil and transfer the potato cake
to a warm plate, to serve.

Potato salad with salmon and dill dressing

SERVES 6

dill, buttermilk and mustard dressing
1 garlic clove, crushed
1^1/$_2$ tablespoons dijon mustard
185 ml (6 fl oz/^{3}/$_4$ cup) buttermilk
185 ml (6 fl oz/^{3}/$_4$ cup) whole-egg
   mayonnaise
3 tablespoons chopped dill

800 g (1 lb 12 oz) all-purpose potatoes,
   such as pontiacs, halved or quartered
800 g (1 lb 12 oz) salmon fillets, skinned
   and bones removed
olive oil, for brushing
200 g (7 oz) watercress sprigs
1 small red onion, very thinly sliced

To make the dressing, combine all the ingredients in a bowl, then whisk until the mixture is smooth. Season to taste with salt and freshly ground black pepper, then cover with plastic wrap and refrigerate until needed.

Cook the potatoes in a saucepan of boiling, salted water for 15–20 minutes, or until tender, then drain well and cool.

Heat a frying pan or barbecue grill over high heat. Brush the salmon all over with olive oil and season with sea salt and freshly ground black pepper. Cook the salmon, in batches if necessary, for 1–2 minutes, or until cooked but still a little pink in the middle. Allow the salmon to cool to room temperature, then break into bite-sized pieces.

Combine the salmon, potatoes, watercress and onion in a large bowl, season to taste and toss to combine well. Divide among bowls or plates, drizzle with the dressing and serve.

Stilton soup

SERVES 4–6

30 g (1 oz) butter
2 leeks, white part only, chopped
1 kg (2 lb 4 oz) potatoes, chopped into
   chunks
1.25 litres (44 fl oz/5 cups) chicken stock
125 ml (4 fl oz/^{1}/$_2$ cup) cream
100 g (3^1/$_2$ oz) stilton cheese
thyme sprigs, to garnish

Melt the butter in a large saucepan, add the leek and cook, stirring often, over medium heat for 6–7 minutes, or until softened. Add the potato and stock, bring to the boil, then reduce the heat and simmer, covered, for 15 minutes, or until the potato is tender.

Allow to cool a little, then transfer the mixture to a blender or food processor and blend or process until smooth.

Return the potato and leek mixture to the saucepan, add the cream and cheese, then stir over low heat until the cheese has melted; do not allow the soup to boil. Divide the soup among warmed bowls, garnish with thyme sprigs and serve immediately.

Potato gnocchi with pancetta and sage

SERVES 4

gnocchi

1 kg (2 lb 4 oz) floury potatoes, unpeeled
2 egg yolks, lightly beaten
2 tablespoons grated parmesan cheese
185 g (6^1/2 oz/1^1/2 cups) plain (all-purpose) flour
extra flour, to knead

sauce

20 g (^{3}/4 oz) butter
80 g (2^3/4 oz) pancetta or bacon slices, cut into thin strips
8 very small sage or basil leaves
150 ml (5 fl oz) thick (double/heavy) cream
50 g (1^3/4 oz/^{1}/2 cup) shaved parmesan cheese

Preheat the oven to 180°C (350°F/Gas 4). Prick the potatoes all over, then put in a baking tin and bake for 1 hour, or until tender. Cool for 15 minutes, then peel and mash, or put through a potato ricer or a mouli (don't use a blender or food processor).

Mix in the egg yolks and parmesan cheese, then gradually stir in the flour. When the mixture gets too firm to use a spoon, work with your hands. Once a loose dough forms, transfer to a lightly floured surface and knead gently. Work in enough extra flour to give a soft, pliable dough that is damp to the touch but not sticky. Divide the dough into six portions. Working with one portion at a time, roll out on the floured surface to make a rope about 1.5 cm (^{5}/8 inch) thick, then cut the rope into 1.5 cm (^{5}/8 inch) lengths. Take one piece of dough and press your finger into it to form a concave shape, then roll the outer surface over the tines of a fork to make deep ridges. Fold the outer lips in towards each other to make a hollow in the middle. Set aside and continue with the remaining dough.

Bring a large saucepan of salted water to the boil. Add the gnocchi in batches, about 20 at a time. Stir gently and cook for 1–2 minutes, or until they rise to the surface. Remove with a slotted spoon, drain and put the gnocchi in a greased shallow casserole dish.

Increase the oven to 200°C (400°F/Gas 6). To make the sauce, melt the butter in a frying pan and fry the pancetta until crisp. Stir in the sage leaves and cream. Season with sea salt and freshly ground black pepper and simmer for 10 minutes, or until thickened.

Pour the sauce over the gnocchi, toss gently and sprinkle the parmesan on top. Bake for 10–15 minutes, or until the cheese melts and turns golden. Serve hot.

Until it was discovered that their vitamin C content was a cure for scurvy, **potatoes** were mainly fed to pigs. For such a poisonous plant (only the tubers are edible), the potato has made a huge impact on the global diet, providing more protein and energy than any other food crop, per unit of land.

pumpkin

Pumpkins are members of the gourd family and are classified as a winter squash. From silky pumpkin soups, creamy pumpkin risottos, sweet pumpkin pies and scones, to the soft, caramelised baked pumpkin alongside the traditional Sunday roast, there could be no more comforting winter vegetable than this.

Buying and storing

- Pumpkin is available year-round, but its peak season is from autumn to winter.
- If buying cut pieces of pumpkin, choose those with bright, deep orange flesh and healthy looking, moist interiors.
- If stored in a cool, well-ventilated place, whole pumpkins will last for several months. They should be unblemished and have thick skin, with an amount of stalk remaining to protect the interior from damp.
- Cut pumpkin is very perishable, so store in the fridge and eat within a few days.

Preparation

Thin-skinned varieties don't need peeling as the skin is edible, although many cooks prefer to remove it first. Thick skin should be cut off with a large sturdy knife (take care doing this, as the skin can be quite hard to cut).

Roasting This cooking method concentrates the pumpkin's sweetness better than any other. Cut the pumpkin into chunks or wedges, toss in olive oil and roast in a 200°C (400°F/Gas 6) oven for 20–30 minutes, or until golden and very soft. Even pumpkin destined for soup benefits from roasting first, as boiling pumpkin in liquid dulls its flavour a little.

Steaming This is a good method for cooking pumpkin if you then want to mash or purée it. Steam large chunks for about 15 minutes, or until soft. To get rid of the moisture (do this if you want to mash it or use it in pumpkin pie), put the steamed pumpkin in a saucepan over medium–high heat for 3–4 minutes, stirring continuously to stop it catching and burning.

Native Americans have been growing and using **pumpkins** for centuries, as have the Mexicans. In Mexico, the flowers are enjoyed as a delicacy, and pumpkin seeds, called pepitas, are roasted and used as a snack, or ground for use as a thickener. The seeds provide oil too, and the shells of some varieties of pumpkin were used to make containers and spoons.

Pumpkin, feta and pine nut pasties

SERVES 4

800 g (1 lb 12 oz) pumpkin (winter squash),
    skin removed and cut into 1 cm (1/2 inch)
    thick slices
2 tablespoons olive oil
3 garlic cloves, crushed
4 sheets ready-made puff pastry, cut into
    15 cm (6 inch) squares
100 g (3 1/2 oz) marinated feta cheese in oil
3 tablespoons oregano, roughly chopped
2 tablespoons pine nuts, toasted
1 egg yolk
1 tablespoon milk
1 tablespoon sesame seeds

Preheat the oven to 220°C (425°F/Gas 7). Put the pumpkin on a baking tray and toss with the combined olive oil and garlic and season with sea salt and freshly ground black pepper. Roast for 40 minutes, or until tender and golden. Remove from the oven and cool.

Evenly divide the pumpkin among the four pastry squares, placing it in the centre. Divide the feta, oregano and pine nuts among the pastry squares, on top of the pumpkin. Drizzle with a little of the feta marinating oil.

Bring two of the diagonally opposite corners together and pinch in the centre above the filling. Bring the other two diagonally opposite corners together and pinch to seal along the edges. The base will be square and the top will form a pyramid. Twist the top to seal where all four corners meet.

Place the egg yolk and milk in a small bowl and whisk with a fork to make an egg wash for the pastry.

Place the pasties on greased baking trays and brush with the egg wash. Sprinkle with the sesame seeds and some sea salt and bake for 15 minutes, or until golden brown.

Tortellini filled with pumpkin and sage

SERVES 6

filling
900 g (2 lb) pumpkin (winter squash),
    peeled and cubed
125 ml (4 fl oz/$^1/_2$ cup) olive oil
1 small red onion, finely chopped
90 g (3$^1/_4$ oz/$^1/_3$ cup) ricotta cheese
1 egg yolk, beaten
25 g (1 oz/$^1/_4$ cup) grated parmesan cheese
1 teaspoon ground nutmeg
2 tablespoons chopped sage

1 packet fresh lasagne sheets
1 egg
2 teaspoons milk
250 g (9 oz) butter
1 handful sage
grated parmesan cheese, to serve

Preheat the oven to 190°C (375°F/Gas 5). To make the filling, put the pumpkin in a roasting tin with half the olive oil and lots of salt and freshly ground black pepper. Bake for 40 minutes, until the pumpkin is very soft.

Meanwhile, heat the remaining oil in a saucepan and gently cook the onion until soft. Put the onion and pumpkin in a bowl, draining off any excess oil. Mash the pumpkin and leave to cool, then crumble in the ricotta. Mix in the egg yolk, parmesan, nutmeg and sage. Season well.

To make the tortellini, cut the lasagne sheets into 8 cm (3$^1/_4$ inch) squares. Mix together the egg and milk to make an egg wash and brush lightly over the pasta just before you fill each one. Put a teaspoon of filling in the middle of each square and fold it over diagonally to make a triangle, pressing down the corners. Pinch together the two corners on the longer side. (If you are not using them immediately, space them out on baking paper dusted with cornmeal and cover with a cloth. Leave for 1–2 hours before cooking — don't refrigerate or they will become damp.)

Cook the tortellini in small batches in a large saucepan of boiling salted water until al dente. Remove and drain with a slotted spoon.

To make the sage butter, melt the butter slowly in a saucepan with the sage and leave to infuse for at least 5 minutes. Drizzle the sage butter over the tortellini and serve with a sprinkling of parmesan.

There are myriad varieties of **pumpkin**, they come in all shapes, colours and sizes, from those the size of tennis balls to some that weigh up to 45 kg (100 lb). Generally, pumpkins have a very similar flavour although sweetness, flavour intensity, water content and colour of the flesh will vary from type to type.

Pumpkin and borlotti bean soup

SERVES 4–6

350 g (12 oz) dried borlotti (cranberry)
    beans
1 kg (2 lb 4 oz) butternut pumpkin (squash),
    peeled, seeded and chopped
2 large all-purpose potatoes, peeled and
    chopped
2 litres (70 fl oz/8 cups) chicken stock
1 tablespoon olive oil
1 red onion, chopped
2 garlic cloves, finely chopped
1 celery stalk, sliced
6–8 sage leaves, chopped
crusty Italian bread, to serve

Put the dried borlotti beans in a large bowl, cover with plenty of cold water and leave to soak overnight.

Rinse the beans well, then put in a saucepan, cover with plenty of fresh cold water and bring to the boil for 5 minutes. Reduce the heat and simmer for 1½ hours, or until tender. Drain well and set aside.

Put the pumpkin, potato and stock in a large saucepan. Bring to the boil, then reduce the heat and simmer for 35–40 minutes, or until the vegetables are soft. Drain well, reserving the liquid. Mash the pumpkin and potatoes, then return to the pan with the reserved liquid. Stir in the beans.

Heat the olive oil in a saucepan. Add the onion, garlic and celery and sauté over medium heat for 6–7 minutes, or until softened. Add to the soup along with the sage and season with freshly ground black pepper and gently heat through. Serve hot, with crusty Italian bread.

Hot pumpkin with taleggio and herbs

SERVES 4–6 AS A SIDE DISH

1 kg (2 lb 4 oz) butternut pumpkin (squash),
    cut into 4 cm (1½ inch) cubes
125 g (4½ oz) taleggio or other washed-rind
    cheese, thinly sliced
1 tablespoon chopped parsley
1 teaspoon chopped oregano
1 teaspoon thyme
1 teaspoon ground nutmeg

Put the pumpkin in a large steamer and cover with a lid. Sit the steamer over a saucepan or wok of simmering water and steam for 15–20 minutes, or until the pumpkin is nearly tender.

Preheat the oven to 200°C (400°F/Gas 6). Transfer the pumpkin to a baking dish and bake for 30 minutes, or until the pumpkin is golden brown. Arrange the cheese on top and bake for a further 3–4 minutes, or until the cheese has melted.

Combine the herbs and nutmeg and sprinkle over the melted cheese. Season with sea salt and freshly ground black pepper, and serve.

Pumpkin pie

SERVES 8

150 g (5 1/2 oz/1 1/4 cups) plain
    (all-purpose) flour
2 teaspoons caster (superfine) sugar
100 g (3 1/2 oz) cold unsalted butter,
    chopped
4 tablespoons iced water
ice cream or whipped cream, to serve
    (optional)

glaze
1 egg yolk
1 tablespoon milk

filling
2 eggs
140 g (5 oz/3/4 cup) soft brown sugar
500 g (1 lb 2 oz) pumpkin (winter squash),
    cooked and cooled, then mashed
4 tablespoons cream
1 tablespoon sweet sherry
1 teaspoon ground cinnamon
1/2 teaspoon ground nutmeg
1/2 teaspoon ground ginger

Lightly grease a 23 cm (9 inch) round pie
dish with butter.

Sift the flour into a large bowl and stir in the
sugar. Using your fingertips, lightly rub in
the butter until the mixture resembles
breadcrumbs. Make a well in the centre, then
add 3 tablespoons of the iced water to the
well. Mix using a flat-bladed knife until a
rough dough forms, adding a little more iced
water if necessary.

Gather the dough together and roll out
between two sheets of baking paper until
large enough to cover the base and side of
the pie dish. Roll the pastry around the rolling
pin, then lift and ease it into the pie dish,

gently pressing to fit the side. Trim away the
excess pastry and crimp the edges.

Roll out the pastry trimmings until 2 mm
(1/8 inch) thick. Using a knife, cut out leaf
shapes of different sizes, then score vein
markings onto the leaves. Cover the
pastry-lined dish and the leaf shapes with
plastic wrap and refrigerate for 30 minutes.

Meanwhile, preheat the oven to 180°C
(350°F/Gas 4).

Prick the base of the chilled pastry case with
a fork. Line with baking paper and half-fill
with baking beads or rice. Place the pastry
leaves on a baking tray lined with baking
paper. To make the glaze, mix together the
egg yolk and milk and brush over the leaves.

Bake the pastry case for 10 minutes, then
remove the paper and baking beads and
bake for a further 10 minutes, or until the
pastry is lightly golden. Meanwhile, bake the
pastry leaves for 10–15 minutes, or until
lightly golden. Remove from the oven and set
aside to cool slightly.

To make the filling, whisk the eggs and sugar
together in a large bowl. Add the remaining
ingredients and stir together well. Pour the
mixture into the pastry case, smoothing over
the surface, then bake for 40 minutes, or
until the filling has set. If the pastry edges
begin to brown too much during cooking,
cover them with foil.

Remove the pumpkin pie from the oven and
leave to cool to room temperature in the dish.
Decorate the top of the pie with the pastry
leaves. Serve with ice cream or whipped
cream, if desired.

Varieties

There are many varieties of sweet potato — their flesh may be white, orange or yellow, ranging from mealy to moist and watery, and their skins may be white, yellow, red, purple or brown. Generally though, they can be loosely divided into three groups.

The orange-skinned sweet potato has orange flesh and is probably the most common. These are softer than the other varieties when cooked. In Australia, this variety is sometimes referred to by its Maori name of kumara. The beige-skinned sweet potato has creamy white flesh, as does the purple-skinned variety.

Buying and storing

- Sweet potatoes are available year-round but are best during autumn and winter.

- Sweet potatoes may be aged for a few weeks before they are sold, as this allows time for their sugars to develop (they are generally not sweet when harvested). Choose ones with smooth, unblemished skins, with no bruises or soft spots.

- Sweet potatoes don't store for long, so only buy as many as you need. Don't store in the fridge; they keep better at room temperature in a well-ventilated spot for 3–5 days.

Preparation

Sweet potatoes are best suited to baking or roasting, as this concentrates their sweet, nutty flavours, or boil them with their skins intact, then peel and use to make mash. They are also delicious when chargrilled — peel and thinly slice them, brush them with oil and cook on a hot chargrill plate for 2–4 minutes on each side. Generally, sweet potatoes can be cooked in much the same way as potatoes, but they take a shorter time to cook.

sweet potato

Not a relation of the potato, as is often thought, and not to be confused with a yam, as it is sometimes mistakenly called, the sweet potato is a member of the morning glory family. Native to the tropical parts of Central and South America, sweet potatoes are still largely associated with the cooking of these areas.

Sweet potato rösti

800 g (1 lb 12 oz) orange sweet potatoes,
    unpeeled
2 teaspoons cornflour (cornstarch)
40 g (1$^1/_2$ oz) butter
150 g (5$^1/_2$ oz) mozzarella cheese, cut into
    30 pieces

Boil the sweet potatoes for 15 minutes, or until almost cooked, but still firm. Set aside to cool, then peel and roughly grate into a bowl. Add the cornflour and $^1/_2$ teaspoon salt and toss lightly to combine.

Preheat the oven to 120°C (235°F/Gas $^1/_2$). Melt some of the butter in a non-stick frying pan over medium heat. Spoon teaspoons of the sweet potato mixture into the pan and put a piece of cheese in the centre of each. Top with another teaspoon of the sweet potato mixture and gently flatten to form rough circles. Cook for 3 minutes on each side, or until golden.

Remove with a slotted spoon and drain on paper towels, then place in the oven to keep warm while cooking the remaining rösti. Serve hot.

note The sweet potato can be cooked and grated up to 2 hours ahead and set aside, covered, until ready to cook.

Split pea and sweet potato soup

SERVES 4

4 tablespoons olive oil
1 large onion, chopped
2 garlic cloves, finely chopped
2 teaspoons finely chopped ginger
110 g (3$^3/_4$ oz/$^1/_2$ cup) yellow split peas
1 red chilli, seeded and sliced
$^1/_2$ teaspoon sweet smoked Spanish paprika
1 litre (35 fl oz/4 cups) chicken stock
500 g (1 lb 2 oz) orange sweet potato,
    cubed
1 handful mint, finely chopped

Heat 1 tablespoon of the olive oil in a large saucepan over medium heat. Fry the onion, garlic and ginger for 4–5 minutes, or until soft and golden. Stir in the split peas, chilli and paprika and cook for 1 minute, or until fragrant. Add the stock and bring to the boil. Reduce the heat and simmer for 20 minutes.

Add the sweet potato, return to the boil, then reduce the heat and simmer for 15 minutes, or until the sweet potato is tender.

Meanwhile, heat the remaining oil in a small saucepan over low heat. Stir in the mint, then immediately remove the saucepan from the heat. Transfer the mint and oil to a small dish.

Remove the soup from the heat. Using a hand-held stick blender fitted with the chopping blade, process for 30 seconds, or until puréed. Season to taste with salt and freshly ground black pepper. Ladle the soup into four bowls and drizzle with a little of the minted oil.

Beef tagine with sweet potato

SERVES 4–6

1 kg (2 lb 4 oz) blade or chuck steak
3 tablespoons olive oil
1 onion, finely chopped
1/2 teaspoon cayenne pepper
1/2 teaspoon ground cumin
1 teaspoon ground turmeric
1/2 teaspoon ground ginger
2 teaspoons paprika
2 tablespoons chopped flat-leaf (Italian)
    parsley
2 tablespoons chopped coriander (cilantro)
2 tomatoes
500 g (1 lb 2 oz) orange sweet potatoes

Trim the steak of any fat and cut into 2.5 cm (1 inch) pieces. Heat half the olive oil in a saucepan and brown the beef in batches over high heat, adding a little more oil as needed. Set aside in a dish.

Reduce the heat to low, add the onion and the remaining oil to the pan and gently cook for 10 minutes, or until the onion is softened. Add the cayenne pepper, cumin, turmeric, ginger and paprika, cook for a few seconds, then add 1 teaspoon of sea salt and freshly ground black pepper. Return the beef to the pan, add the parsley, coriander and 250 ml (9 fl oz/1 cup) water. Cover and simmer over low heat for 1 1/2 hours, or until the meat is almost tender.

Peel the tomatoes. To do this, score a cross in the base of each one using a knife. Put the tomatoes in a bowl of boiling water for 20 seconds, then plunge them into iced water to cool. Remove from the water and peel the skin away from the cross — the skins should slip off easily. Slice the tomatoes. Peel the sweet potatoes, cut them into 2 cm (3/4 inch) dice and leave them in a bowl of water until required.

Preheat the oven to 180°C (350°F/Gas 4). Transfer the meat and its sauce to an ovenproof serving dish (the base of a tagine would be ideal). Drain the sweet potato and spread it on top of the beef. Top with the tomato slices. Cover with foil (or the lid of the tagine) and bake for 40 minutes. Remove the foil, increase the oven temperature to 220°C (425°F/Gas 7) and raise the dish to the upper oven shelf. Cook until the tomato and sweet potato are flecked with brown and are tender. Serve immediately.

Sweet potatoes can be cooked as for regular potatoes. However, when frying sweet potatoes, their high sugar content means they can easily burn, so fry them at a lower temperature than you would for potatoes.

Sweet potato filo pie

SERVES 8

750 g (1 lb 10 oz) orange sweet potato,
    peeled
12 small French shallots, peeled
6 small potatoes, peeled and halved
125 ml (4 fl oz/^{1}/2 cup) olive oil
1 teaspoon sweet paprika
1 teaspoon ground ginger
2 teaspoons ground cumin
^{1}/4 teaspoon ground cinnamon
100 g (3^1/2 oz/2 cups) baby English spinach
60 g (2^1/4 oz/^{1}/2 cup) sultanas (golden
    raisins)
85 g (3 oz/^{2}/3 cup) slivered almonds,
    toasted
100 g (3^1/2 oz/^{2}/3 cup) pistachio kernels,
    roughly chopped
1 large handful coriander (cilantro),
    chopped
2^1/2 tablespoons golden syrup or
    dark corn syrup
80 g (2^3/4 oz/^{1}/3 cup) plain yoghurt
400 g (14 oz) tin chickpeas, rinsed and
    drained
3 garlic cloves, finely chopped
a pinch of cayenne pepper
3 tablespoons lemon juice
125 g (4^1/2 oz) butter, melted
9 sheets ready-made filo pastry

Preheat the oven to 200°C (400°F/Gas 6).
Cut the sweet potato into 2.5 cm (1 inch)
cubes and put in a large roasting tin, along
with the shallots and potatoes. Combine the
olive oil, paprika, ginger, cumin and cinnamon
in a small bowl and pour over the vegetables.
Toss to coat. Transfer to the oven and roast
for 25 minutes, then turn the vegetables and
roast for a further 15 minutes. Remove from
the oven and reduce the oven temperature
to 180°C (350°F/Gas 4).

Add the English spinach and sultanas to the
vegetables. Toss lightly, then set aside for
5 minutes for the spinach to wilt. Transfer
the vegetables to a large bowl and add the
almonds, pistachios and coriander.

Put 2 tablespoons of the golden syrup, the
yoghurt, chickpeas, garlic, cayenne pepper
and lemon juice in a food processor and
blend until smooth. Season with salt and
freshly ground black pepper. Add to the
vegetables and mix through.

Brush a 28 x 21 cm (11^1/4 x 8^1/4 inch)
loose-based rectangular tart tin with butter.
Brush a sheet of filo pastry with butter and
lay it diagonally over the tin, so that three of
the points stick out. Don't push the pastry
into the sides of the tin, just place it loosely
on top. Brush another sheet of filo with
butter and lay it similarly, at the opposite end
of the tin. Brush a third sheet with butter and
lay it in the middle of the tin. Continue in this
way twice more, until all the filo is used.

Pile the sweet potato mixture in the centre
of the tin. Starting in the middle, bring the
opposite sides of the filo together, encasing
the filling tightly but with the filo points
sticking up. Brush the filo carefully with the
remaining butter and drizzle the remaining
golden syrup in zigzags over the top.

Place the tin on a baking tray and transfer
to the oven. Bake for 30 minutes, or until
the pastry is golden. Set aside for 5 minutes
before serving.

turnip & swede

Closely related, these wonderfully versatile vegetables can be added to stews and casseroles; mash them with potatoes; or purée them into creamy soups. While unpopular in some countries, they are loved in others: the Scots traditionally serve their haggis with 'bashed neeps' — mashed buttered swedes.

Buying and storing

- Turnips and swedes (also called rutabagas) are available year-round in markets, although they are cultivated during late autumn and winter, and through to spring. Winter turnips can have tough, thick skins and be on the large side but their flavour will still be good; it's not worth buying turnips in summer.

- Try to buy turnips with their green tops still intact, as this indicates they have recently been harvested, and the same applies for swedes. Avoid any with shrivelled or damaged skins.

- When buying swedes, choose those with a purplish top and fresh green stalks for the finest flavour. Choose those that are heavy for their size, as these will be juicy.

- Turnips don't store for long and will quickly turn bitter, so store in a plastic bag in the fridge and use within a few days.

- Swedes will last for 3–4 weeks in the fridge, though they will soften and their flavour will strengthen. Should you find the flavour too strong, simply blanch them for 10 minutes, then discard the cooking water and resume cooking in fresh water.

Preparation

Before cooking, both swedes and turnips need to be trimmed and peeled. If the flesh just under the skin is tough and fibrous, you will need to pare this away too.

Cut turnips into large chunks and boil in salted water for 15–20 minutes. Swedes can be cooked in the same way as turnips, but will take about 25 minutes to cook — never overcook them, as this ruins both their flavour and texture. Both are delicious steamed until tender, then tossed with burnt butter or fruity olive oil.

Spiced baby turnips

SERVES 4

400 g (14 oz) small roma (plum) tomatoes
3 tablespoons olive oil
3 small onions, sliced
3 teaspoons ground coriander
1 teaspoon sweet paprika
350 g (12 oz) baby turnips, trimmed
1 teaspoon soft brown sugar
600 g (1 lb 5 oz/$^{1}/_{2}$ bunch) silverbeet
   (Swiss chard)
1 handful parsley

Peel the tomatoes. To do this, score a cross in the base of each one using a knife. Put the tomatoes in a bowl of boiling water for 20 seconds, then plunge them into iced water to cool. Remove from the water and peel the skin away from the cross — the skins should slip off easily. Cut the tomatoes widthways into 1.5 cm ($^{5}/_{8}$ inch) slices and squeeze out most of the juice and seeds.

Heat the olive oil in a large frying pan and fry the onions for 5–6 minutes over medium heat, or until soft. Stir in the coriander and paprika, cook for 1 minute, then add the tomato, turnips, sugar and 4 tablespoons hot water. Season well. Cook over medium heat for 5 minutes, then cover the pan, reduce the heat to low and cook for 4–5 minutes, or until the turnips are tender.

Meanwhile, strip the silverbeet leaves off the stalks. Rinse under cold water and shake off the excess.

Stir the parsley and silverbeet into the pan with the turnips, check the seasoning and cook, covered, for 4 minutes, or until the silverbeet is wilted. Serve hot.

Baked swedes with blue cheese and ricotta

SERVES 4

4 swedes (rutabagas), scrubbed
1$^{1}/_{2}$ tablespoons olive oil
40 g (1$^{1}/_{2}$ oz) butter
55 g (2 oz) creamy blue vein cheese,
   crumbled
125 g (4$^{1}/_{2}$ oz/$^{1}/_{2}$ cup) ricotta cheese
   (from the delicatessen)
12 sage leaves
1 garlic clove, crushed

Preheat the oven to 180°C (350°F/Gas 4). Rub the swedes with a third of the olive oil. Place each swede in the centre of a 30 cm (12 inch) square of foil, season lightly with salt and freshly ground black pepper and dot with the butter. Fold up the foil to enclose the swedes. Arrange, root down, in a small baking tin and bake for 1 hour, or until tender.

Combine the blue cheese and ricotta in a small saucepan and heat over low heat until soft. Keep warm. Heat the remaining olive oil in a small frying pan until hot. Fry the sage leaves for a few seconds until crisp, then remove and drain on paper towels.

Add the garlic to the pan, reduce the heat to low and fry until the garlic just begins to colour. Transfer to the cheese mixture and add 4 of the fried sage leaves. Season to taste with salt and freshly ground black pepper and stir to combine.

Remove the foil from the swedes and cut them diagonally into 2–3 cm ($^{3}/_{4}$–1$^{1}/_{4}$ inch) slices. Reassemble to serve, with the sauce spooned over the top. Top with the remaining sage leaves.

Finnish creamy baked swede

SERVES 6–8 AS A SIDE DISH

1.6 kg (3 lb 8 oz) swedes (rutabagas)
(about 4), peeled and cut into 4 cm
(1$\frac{1}{2}$ inch) pieces
125 ml (4 fl oz/$\frac{1}{2}$ cup) cream
2 eggs, lightly beaten
1 egg yolk
3 tablespoons plain (all-purpose) flour
$\frac{1}{2}$ teaspoon ground nutmeg
a small pinch of ground cloves
100 g (3$\frac{1}{2}$ oz/1$\frac{1}{4}$ cups) fresh breadcrumbs
50 g (1$\frac{3}{4}$ oz) unsalted butter, chopped
4 sage leaves, finely chopped

Cook the swedes in boiling salted water for 40 minutes, or steam them for 25 minutes, until tender. Drain well and return to the saucepan, then mash using a potato masher — the mixture should still have some texture.

Meanwhile, preheat the oven to 180°C (350°F/Gas 4). Grease a shallow 18 x 30 cm (7 x 12 inch) baking dish.

Return the saucepan to the stovetop over medium heat and cook the swedes, stirring often, for 5–7 minutes, or until the excess liquid has evaporated. Remove from the heat and allow to cool slightly.

Stir in the cream, eggs, egg yolk, flour, nutmeg and cloves, then season with sea salt and freshly ground black pepper. Pour the mixture into the baking dish, smoothing the top so it is even.

Put the breadcrumbs and butter in a food processor and process until fine clumps form, then stir in the sage. Sprinkle the mixture evenly over the swede. Transfer to the oven and bake for 30–35 minutes, or until golden and set in the middle. Serve hot or warm as a side dish to accompany roasted or grilled meats.

variation Substitute the same weight of turnips for swedes in this recipe, if preferred.

The name '**swede**' comes from the fact that this crop is common in Sweden; some claim it is native to that country although its origins are perhaps Siberian or Finnish. In America, swedes are commonly called rutabagas, derived from the Swedish, *rotabagge*, meaning 'root ram'. And just to confuse things even more, the plant is sometimes referred to there as a yellow turnip or wax turnip.

the onion family

garlic

Garlic is the most widely used herb flavouring in the culinary world — imagine garlic prawns, pesto, garlic bread and chicken with forty cloves of garlic without the sweet pungency of garlic. For a long time garlic was used only medicinally or was considered a food for the poor, but gradually it has found its place as an essential staple in every good cook's larder.

Buying and storing

- Garlic is available year-round, but its peak season is during spring and summer.

- There are many varieties of garlic; they all taste essentially the same, they just vary in intensity of flavour. When buying garlic, choose bulbs with plump, firm cloves and a fat neck, as these have a more delicate flavour. The papery shells should be crisp and dry and tight around the cloves.

- Keep garlic away from humidity in a cool, dark place, and it will keep for a few weeks. Garlic should not be frozen, as this ruins both flavour and texture.

Preparation

Peeling The easiest way to peel garlic is to lay the cloves on a board, place the flat side of a large knife blade over the clove and bang the knife with the heel of your hand — the skin will then just pull off. Don't prepare garlic until you need it, as exposure to the air causes bitterness.

Chopping The best way to chop quantities of garlic is to chop the cloves with a little salt, which will help to break down and soften the garlic. For convenience, you can use a garlic crusher to crush your cloves, although some cooks maintain that this alters the flavour a little.

The volatile component in garlic responsible for its smell is allinaise, which is released when the cloves are cut or crushed — the more finely the garlic is cut, the more allinaise is released. Heat destroys allinaise, so the longer you cook garlic, the milder the flavour (take care, however, to cook it over low–medium heat, as overbrowned garlic can become bitter). In the French dish, chicken with forty cloves of garlic, the garlic flavour is actually mild because the garlic is cooked whole and for some time. However, just one clove of raw crushed garlic added to a salad dressing will be quite potent.

Garlic prawns

SERVES 4

25 g (1 oz) butter
100 ml (3 1/2 fl oz) olive oil
2 garlic cloves, finely chopped
1 small red chilli, seeded and finely chopped
24 large raw prawns (shrimp), peeled and
    deveined, tails intact
3 tablespoons chopped flat-leaf (Italian)
    parsley
lemon wedges, to serve
crusty bread, to serve

Heat the butter and olive oil together in a large frying pan. When hot, add the garlic and chilli. Cook, stirring continuously, for 30 seconds.

Add the prawns to the pan and cook for 3–4 minutes on each side, or until the prawns turn pink. Sprinkle with the parsley and serve immediately on hot plates, with lemon wedges and crusty bread.

Aïoli

MAKES ABOUT 500 G (1 LB 2 OZ/2 CUPS)

4 egg yolks
8 garlic cloves, crushed
2 tablespoons lemon juice
500 ml (17 fl oz/2 cups) olive oil

Put the egg yolks, garlic, half the lemon juice and 1/2 teaspoon salt in a mortar or food processor and pound with a pestle or process until light and creamy.

Add the olive oil, drop by drop from the tip of a teaspoon, whisking constantly until it begins to thicken, then add the oil in a very thin stream. (If you're using a food processor, pour in the oil in a thin stream with the motor running.) Season with salt and freshly ground black pepper. Add the remaining lemon juice and, if necessary, thin with a little warm water.

Serve the aïoli as a dipping sauce. Keep aïoli sealed in a sterilised jar in the fridge for up to 3 weeks. Serve with poached fish, chicken or seafood, or with roast beef fillet or as an accompaniment to whole baked potatoes.

All sorts of powers have been attributed to **garlic**, from warding off demons and vampires, to its use as a remedy for the common cold. The Egyptian labourers building the pyramids were given garlic to make them stronger, Eleanor Roosevelt ate chocolate-coated garlic pills every morning to help her memory, and Pliny the Elder took a garlic, spice and wine love potion to improve his manly vigour.

leek

A mild-flavoured member of the onion family, the thick white stems of cultivated leeks are blanched by piling dirt up around them as they grow, which explains why they are notoriously dirty and require careful cleaning. Like onions, leeks need to be cooked for a long time to render them soft and sweet.

Buying and storing

- Leeks are available year-round, however their peak season is from late spring to mid-winter.

- Buy leeks that feel firm and are heavy for their size, with dark green leaves and no yellowing. Smaller, thinner leeks tend to be less fibrous and sweeter than the larger ones.

- Store leeks, with leaves attached, in the fridge in a loosely sealed plastic bag for up to 7 days.

Preparation

The edible part of the leek is the white part and the palest part of the green tops only; the coarse green tops need to be cut off and discarded (or wash them and use in stocks).

Leeks hold lots of dirt between their layers, so wash them well before use. If you plan to use them sliced or cut into matchsticks, the cut leeks can be washed in water to remove any soil. If you want to serve them whole, make a deep slit through the leafy end of the leeks and into the white part, then soak them in water, shaking them well to dislodge any dirt, and drain them leafy side down. Alternatively, cut the leeks in half lengthways, keeping them attached at the root, and rinse them well under running water.

Leeks can be braised whole in chicken stock, with some herbs such as thyme. They are also excellent stewed — cooked some chopped leeks in butter and finish with some cream and a pinch of nutmeg. They can be steamed too: cut them into 5 cm (2 inch) rounds and steam for 10–15 minutes, or trim and steam whole leeks for about 20 minutes. Never simply boil leeks, as they will be soggy and tasteless.

Chicken and leek pie

SERVES 4

50 g (1 3/4 oz) butter
2 large leeks, white part only, thinly sliced
4 spring onions (scallions), sliced
1 garlic clove, crushed
30 g (1 oz/1/4 cup) plain (all-purpose) flour
375 ml (13 fl oz/1 1/2 cups) chicken stock
125 ml (4 fl oz/1/2 cup) cream
1 barbecued chicken, meat removed from
    bones and chopped
375 g (13 oz) block ready-made puff pastry
3 tablespoons milk

Preheat the oven to 200°C (400°F/Gas 6).
Grease a 20 cm (8 inch) round pie tin.

Melt the butter in a saucepan and add the
leeks, spring onions and garlic. Cook over low
heat for 6 minutes, or until the leeks are soft
but not browned. Sprinkle with the flour and
mix well. Gradually pour in the stock and
cook, stirring, until the mixture is thick and
smooth. Stir in the cream and add the
chicken. Put the mixture in the pie tin and
set aside to cool.

Cut the pastry in half and roll out each piece
between two sheets of baking paper to a
thickness of 3 mm (1/8 inch). Cut a circle out
of one of the sheets of pastry to cover the
top of the pie. Brush the rim of the pie tin
with a little milk. Put the pastry on top and
press to seal around the edge. Trim off any
excess pastry and crimp the edge with the
back of a fork.

Cut the other sheet into 1 cm (1/2 inch) strips
and roll each strip up loosely like a snail.
Arrange on top of the pie, starting from the
centre and leaving a gap between each one.
The spirals may not cover the whole surface
of the pie. Make a few small holes between
the spirals to let out any steam, and brush
the pastry spirals lightly with milk.

Bake for 25–30 minutes, or until the top is
brown and crispy. Make sure the spirals look
well cooked and are not raw in the middle.
Serve immediately.

note Make individual chicken and leek pies
by dividing the mixture among four greased
310 ml (10 fl oz/1 1/4 cup) round ovenproof
dishes. Cut the pastry into 4 rounds to fit.
Bake for 15 minutes, or until the pastry is
light brown and crisp.

Like the other members of the allium family, including
shallots, onions and garlic, **leeks** have great health-
promoting properties. They are attributed with lowering
high cholesterol levels, protecting against high blood
pressure and also against some cancers.

Leeks à la grecque

SERVES 4 AS A SIDE DISH OR STARTER

3 tablespoons extra virgin olive oil
1¹/₂ tablespoons white wine
1 tablespoon tomato paste (concentrated
    purée)
¹/₄ teaspoon sugar
1 bay leaf
1 thyme sprig
1 garlic clove, crushed
4 coriander seeds, crushed
4 peppercorns
8 small leeks, white part only, rinsed well
1 teaspoon lemon juice
1 tablespoon chopped parsley
lemon halves or wedges, to serve

Put the olive oil, wine, tomato paste, sugar, bay leaf, thyme, garlic, coriander seeds, peppercorns and 250 ml (9 fl oz/1 cup) water in a large heavy-based frying pan with a lid. Bring to the boil, cover and simmer for 5 minutes.

Add the leeks in a single layer and bring to simmering point. Reduce the heat, then cover and simmer gently for 20–30 minutes, or until the leeks are tender when pierced with a skewer. Drain the leeks well, reserving the liquid, then transfer to a serving dish.

Add the lemon juice to the reserved cooking liquid and boil rapidly for 1 minute, or until the liquid has reduced and is slightly syrupy. Season to taste with sea salt, then strain the sauce over the leeks.

Allow to cool, then serve the leeks at room temperature, sprinkled with chopped parsley and with some lemon, for squeezing over.

Braised leek with pine nuts

SERVES 4 AS A SIDE DISH

20 g (³/₄ oz) butter
2 teaspoons olive oil
2 leeks, white part only, thinly sliced
4 tablespoons vegetable stock
4 tablespoons dry white wine
2 tablespoons finely chopped mixed herbs,
    such as flat-leaf (Italian) parsley and
    oregano
2¹/₂ tablespoons pine nuts, lightly toasted
4 tablespoons grated parmesan cheese

Heat the butter and olive oil in a large frying pan. Add the leeks and sauté for 5 minutes, or until golden brown.

Add the stock and wine and cook for a further 10 minutes, or until the leeks are tender. Stir in the herbs, sprinkle with the pine nuts and parmesan. Serve as a side dish with roast meats or grilled fish.

onion

Onions weren't always popular —
they were once considered fit only
for the poor: they were cheap and
easy to grow and were frequently
eaten raw on bread. Today they
are rightfully considered one of
the most important ingredients
in the kitchen, they are used in
just about every nation's cuisine,
adding a depth of flavour to many
savoury recipes, and they are also
a wonderful vegetable in their own
right. Onions are sold as either dry
onions or green onions. Dry onions
are left in the ground to mature,
where they develop a papery skin,
whereas green onions, or spring
onions (page 82), are pulled out
while the onion is young and the
bulb is still small.

Varieties

There are over 300 varieties of onions; very few of these are sold by named variety but are generally sold by type. Some of the common dry onions are listed below.

Baby Sometimes called pickling or pearl onions, these very small onions are usually brown-skinned but sometimes white. They can be pickled whole or may be roasted, braised or used in stews such as coq au vin.

Brown These all-purpose onions have either brown or yellow skins, and come in a variety of sizes. Use in stews and for frying, roasting and stuffing. They have quite a strong flavour, which becomes milder and sweeter when cooked. They are not a good onion for eating raw.

Red These reddish-skinned onions have a sweeter flavour than brown onions, making them a good choice for eating raw in salads, dressings, salsas and sandwiches. They are also delicious when chargrilled or barbecued.

Spanish These round, yellow-skinned mild-tasting onions are good for eating raw in salads. Confusingly, in Australia, red onions are also called Spanish onions.

White As the name suggests, these have white skin and white juicy flesh. Their flavour tends to be stronger than brown onions, and they are a good choice for sautéing, roasting, baking and for use in casseroles. White onions contain a fair amount of pungent sulphur compounds, so be mindful of this when peeling them.

Buying and storing

- Dry onions are available year-round.

- Onions should feel firm and heavy, their skin should be shiny and the skin around the neck should be very tightly closed. There should be no soft or dark spots or sprouting. Don't buy onions with green areas, as these will taste unpleasant.

- Unpeeled onions should only have a mild smell — a strong odour can indicate spoilage.

- Store onions in a dark and well-ventilated place such as the pantry, as exposure to light can cause bitterness.

- Onions readily absorb moisture so don't store them under the sink, or near potatoes, as these give off moisture and gas and will make the onions spoil more rapidly.

Preparation

Onions contain volatile sulphur compounds in their cells, which are released when the onion is chopped or sliced, resulting in the inevitable onion tears. The older the onion, the stronger these compounds are.

Peeling To peel a quantity of onions, especially baby ones that will be used whole, pour boiling water over them in a bowl and set aside for 5 minutes to allow the skins to soften and loosen. Whether using this method or not, peel the onions from the neck end down, leaving the root intact (the root will hold the onion together, making it easier for chopping and will keep the onion together if cooking it whole).

Cutting There are many suggestions for reducing the effects of onion-chopping. Try one of these steps to overcome the discomfort: soak the onions in water for 30 minutes or chill them before peeling; peel them under an open window or exhaust fan; or breathe with your mouth open. Use a very sharp knife, as this will give a cleaner cut and therefore reduce the exposure to the sulphur compounds. To chop an onion, first peel and halve it, then place it, flat side down, on a board before slicing and chopping. Do not prepare onions in advance as they quickly oxidise and can taste unpleasant.

Roasting To roast onions, either peel them or leave them unpeeled. When unpeeled, the onion flesh steams to a creamy softness inside the skins, and peeled roast onions result in a crusty, caramelised exterior and a softish interior. Either way, trim a little off the base of each onion so they will stand up in the dish, rub generously with oil, then roast in a 180°C (350°F/Gas 4) oven for about 1 hour, or until tender.

Chargrilling Cut peeled onions into thin wedges, brush with oil and cook on a barbecue grill or chargrill pan on both sides until lightly charred and tender. These can be drizzled with balsamic vinegar, sprinkled with sea salt and chopped mint and served as a side dish.

Caramelising Onions lose much of their water content during cooking, so start with 6–8 onions to make it worth your while. Thinly slice the onions and put in a saucepan with some olive oil. Cook over medium heat for 45–60 minutes, stirring often, or until golden and very soft. Add some brown sugar or balsamic vinegar near the end of cooking if you like — the onions will be thick and 'jammy'. Serve with roast meats or with mashed potatoes and sausages.

After handling **onions** and cooking with them, you will have a strong aroma left on your hands. Try rubbing them with lemon juice or vinegar, or mix a little olive oil with some salt and lemon juice and use that as a hand cleanser, then rinse well.

Sweet and sour onions

SERVES 6 AS A SIDE DISH

3 red onions (about 500 g/1 lb 2 oz)
2 tablespoons wholegrain mustard
2 tablespoons honey
2 tablespoons red wine vinegar
2 tablespoons olive oil

Preheat the oven to 220°C (425°F/Gas 7).

Carefully peel the onions, keeping the root ends intact so that the layers stay together. Cut each onion lengthways into eight pieces and place in a non-stick baking dish.

Whisk together the remaining ingredients, then drizzle over the onions. Cover and bake for 20 minutes, then remove the lid and bake for a further 15–20 minutes, or until the onions are soft and caramelised. Serve hot.

Onion and thyme marmalade

FILLS THREE 250 ML (9 FL OZ/1 CUP) JARS

2 kg (4 lb 8 oz) onions, cut into rings
750 ml (26 fl oz/3 cups) red or white
    wine vinegar
6 black peppercorns
2 bay leaves
10 cm (4 inch) square of muslin
    (cheesecloth)
800 g (1 lb 12 oz/4$\frac{1}{3}$ cups) soft brown
    sugar
2 tablespoons thyme
10 thyme sprigs, 3 cm (1$\frac{1}{4}$ inches) long

Put the onions in a large saucepan with the vinegar. Put the peppercorns and bay leaves in the muslin and tie securely with string, then add to the saucepan. Bring to the boil, then reduce the heat and simmer for 45 minutes, or until the onions are very soft.

Add the sugar, thyme leaves and 1 teaspoon salt, stirring for 7–8 minutes, or until all the sugar has dissolved. Bring to the boil, then simmer for 20–30 minutes, or until thick and syrupy. Remove any scum from the surface with a slotted spoon. Discard the muslin bag and stir in the thyme sprigs.

Pour into hot sterilised jars and seal. Allow to cool, then label and date each jar. Store in a cool, dark place for 6–12 months. After opening, store in the fridge for up to 6 weeks.

Serve the onion and thyme marmalade with roast beef, venison, sausages and sharp cheeses or as a sandwich relish.

French onion soup

50 g (1 3/4 oz) butter
1 tablespoon olive oil
1 kg (2 lb 4 oz) onions, thinly sliced into rings
3 x 425 g (15 oz) tins chicken or beef
   consommé
125 ml (4 fl oz/1/2 cup) dry sherry
half a baguette
125 g (4 1/2 oz/1 cup) finely grated cheddar
   or gruyère cheese

Heat the butter and oil in a large saucepan, add the onions and cook, stirring frequently, over low heat for 45 minutes, or until the onions are softened and translucent. It is important not to rush this stage — cook the onions thoroughly so that they caramelise and the flavour develops.

Add the consommé to the pan along with the sherry and 250 ml (9 fl oz/1 cup) water. Bring to the boil, then reduce the heat and simmer for 30 minutes. Season to taste.

Meanwhile, slice the baguette into four thick slices and arrange them in a single layer under a hot grill (broiler). Toast one side, then remove from the grill, turn over and cover the untoasted side with the cheese.

Ladle the hot soup into four serving bowls, top each with a slice of toast, cheese side up, and place under the grill until the cheese is melted and golden. Serve immediately.

Roasted red onions and tomatoes

SERVES 4

8 roma (plum) tomatoes
2 red onions
2 garlic cloves
1 1/2 tablespoons balsamic vinegar
1 teaspoon dijon mustard
3 tablespoons extra virgin olive oil

Preheat the oven to 150°C (300°F/Gas 2) and lightly grease a baking tray with oil.

Cut the tomatoes into quarters and put on the tray. Remove the tops of the onions and peel. Cut each onion into 8 wedges and place on the tray with the tomatoes. Place the garlic in the middle of the tray and season the vegetables well. Roast for 1 hour.

Arrange the tomatoes and onions on a serving plate. Peel the roasted garlic and crush it in a small bowl. Add the balsamic vinegar and mustard and beat in the olive oil using a small whisk, adding the oil slowly in a thin stream. Season well and drizzle the dressing over the onions and tomatoes.

shallot

Shallots, or eschalotes, grow in clusters and are joined with a common root end. Prized for their distinct but polite flavours, shallots never 'take over' a dish in the way onions or garlic can, and are valued by cooks the world over, in sauces or cooked whole as a vegetable. In some countries, spring onions are erroneously called shallots.

Varieties

Another member of the allium family, true shallots grow not from seeds but from a bulb. A 'mother' bulb is planted and this produces many 'daughters'.

There are several varieties of shallots and these include the grey or common shallot; the Jersey shallot, a round bulb with pink skin; the French shallot, which has copper-coloured skin and an elongated bulb; and red Asian shallots, which are a light pink colour.

Buying and storing

- Shallots are available year-round.

- Choose bulbs that feel hard and have papery skins that are shiny and brittle. Avoid those that are soft or withered.

- If very fresh when you bought them, shallots will store for several months in a cool, dark, well-ventilated place.

- If you do refrigerate them, wrap them first in paper towels to protect them from humidity.

Preparation

Shallots are generally used thinly sliced or chopped, but they can also be cooked whole. If using a large quantity of shallots, it can be quite time-consuming to peel them. To make this easier, put the shallots in a bowl, cover with boiling water (put a plate on top of them to keep them submerged) and set aside for 5 minutes. Drain and cool slightly.

To peel, slice across the skin from the very top of each bulb, then use a small knife to pull a section of skin away from flesh, then pull it down to the root end. Trim the root ends but don't cut them off completely, as this will help the shallot hold together.

Shallot, bacon and cheddar muffins

MAKES 6

5 French shallots, peeled
3 tablespoons oil
2 bacon slices, finely chopped
250 g (9 oz/2 cups) plain (all-purpose) flour
1 tablespoon baking powder
1 tablespoon caster (superfine) sugar
1 teaspoon dry mustard
140 g (5 oz/scant 1 1/4 cups) grated mature cheddar cheese
185 ml (6 fl oz/3/4 cup) milk
1 egg
sweet paprika, to serve

Preheat the oven to 200°C (400°F/Gas 6). Grease a 6-hole giant muffin tin. Slice 1 of the shallots into rings. Heat 2 teaspoons of the oil in a frying pan over low heat. Add the shallot and fry for 3 minutes. Remove and drain on paper towels. Set aside.

Finely chop the remaining 4 shallots. Increase the heat to medium and add the chopped shallots and bacon. Fry for 5 minutes, or until the shallots are soft. Drain on paper towels.

Sift the flour, baking powder, sugar, mustard and 1/2 teaspoon salt into a bowl. Add 90 g (3 1/4 oz/3/4 cup) of the cheddar cheese, the bacon and shallots and stir through. Combine the milk, egg and remaining oil in a jug. Pour into the bowl and fold gently — the batter should be lumpy.

Divide the batter among the muffin holes. Top with the fried shallot and remaining cheddar cheese. Bake for 20–25 minutes, or until the muffins are golden and cooked through. Cool in the tin before turning out on a wire rack. Sprinkle paprika on top to serve.

Roasted shallot and beet salad

SERVES 4

2 tablespoons red wine vinegar
4 tablespoons walnut oil
1 garlic clove, crushed
1 teaspoon dijon mustard
6 beetroot (beets), plus 70 g (2 1/2 oz) small beetroot leaves
12 French shallots, unpeeled
12 garlic cloves, unpeeled
1 tablespoon vegetable oil
50 g (1 3/4 oz/1/2 cup) walnuts, toasted

Preheat the oven to 200°C (400°F/Gas 6). In a small bowl, whisk together the vinegar, walnut oil, garlic and mustard. Season well and set aside.

Cut the leaves off the beetroot bulbs, leaving 3 cm (1 1/4 inches) of stalk attached, but don't trim the roots. Wash well to remove all traces of soil, then put in a roasting tin along with the shallots, garlic cloves and vegetable oil. Roast in the oven for 1 hour, then remove the shallots and garlic. Continue to roast the beetroot for 30 minutes, or until tender.

Slip the skins off the shallots, garlic cloves and beetroot, then cut the beetroot into wedges. Add the dressing and toss to combine. Cool to room temperature.

Place the beetroot leaves, walnuts and vegetables in a bowl, season well with sea salt and freshly ground black pepper and gently toss together. Arrange on a serving platter or individual plates.

French shallot tatin

SERVES 6

750 g (1 lb 10 oz) large French shallots,
    unpeeled
50 g (1 ³/₄ oz) butter
2 tablespoons olive oil
4 tablespoons soft brown sugar
3 tablespoons balsamic vinegar

pastry
125 g (4 ¹/₂ oz/1 cup) plain (all-purpose)
    flour
60 g (2 ¹/₄ oz) chilled butter, chopped
2 teaspoons wholegrain mustard
1 egg yolk, mixed with 1 tablespoon
    iced water

Put the shallots in a saucepan of boiling water for 5 minutes to make them easier to peel. Drain well, allow to cool slightly, then peel the shallots, taking care to leave the root ends intact.

Heat the butter and olive oil in a large heavy-based frying pan. Add the shallots and cook over low heat, stirring often, for 15 minutes, or until the shallots have started to soften. Add the sugar, balsamic vinegar and 3 tablespoons water and stir to dissolve the sugar. Simmer over low heat for a further 15–20 minutes, or until the liquid has reduced and has become syrupy, turning the shallots occasionally.

To make the pastry, sift the flour and a pinch of salt into a large bowl. Using your fingertips, lightly rub the butter and mustard into the flour until the mixture resembles coarse breadcrumbs. Make a well in the centre. Add the egg yolk mixture to the well and mix using a flat-bladed knife until a dough forms.

Gently gather the dough together, transfer to a lightly floured surface, then press into a round disc. Cover with plastic wrap and refrigerate for 30 minutes, or until firm.

Meanwhile, preheat the oven to 200°C (400°F/Gas 6). Grease a shallow 20 cm (8 inch) round cake tin with butter. Pack the shallots tightly into the tin and drizzle with any syrup remaining in the frying pan.

On a sheet of baking paper, roll out the pastry to a circle 1 cm (¹/₂ inch) larger than the tin. Lift the pastry into the tin and lightly push it down so it is slightly moulded over the shallots. Bake for 20–25 minutes, or until the pastry is golden brown.

Remove the tart from the oven and place the tin on a wire rack for 5 minutes. To serve, place a plate over the tin and carefully turn the tart out onto the plate. Serve warm.

Buying and storing

- Spring onions are available year-round but their peak season is from spring to summer.

- Choose bunches of spring onions with firm white root ends and deep green tops, with no sign of yellowing. The thinner onions will have a milder flavour.

- Store spring onions, covered in plastic wrap, in the crisper of the fridge for up to 3 days.

Preparation

Before cooking, slice off the root end and peel off the fine membrane that covers the white part of the onion. The spring onion can then be thinly sliced on the diagonal (cut off the top 4 cm/1½ inches of green leaves) and added to stir-fries or soups. For garnishing, cut into short lengths, then cut these into fine matchsticks and use to garnish fish, noodle dishes or soups.

spring onion

This onion has a confusing number of names — green onion, shallot and scallion — depending on where you live or what term you prefer. Spring onions are an immature onion that, if left in the ground, would grow to full size. Depending on when it is picked, it has a small, white bulb of varying size, and long green tops. Spring onions have a mild, delicate flavour and are used around the world, particularly in Asian cooking, and are the only onion commonly used in traditional Japanese cooking. Both the green tops and white stems are edible and these can be sliced and added to salads or omelettes, tossed into stir-fries or finely shredded and used as a garnish.

Spring onion mash

SERVES 4–6

1 kg (2 lb 4 oz) floury potatoes
40 g (1 1/2 oz) butter
3 tablespoons hot milk
4 tablespoons hot cream
3 spring onions (scallions), thinly sliced

Peel the potatoes and cut them into large chunks. Cook the potato in boiling salted water for 12 minutes, or until tender, then drain the water and briefly return the potato to the heat, shaking the pan, to evaporate any excess moisture.

Add the butter, hot milk and hot cream, and mash the potato until smooth and lump-free. Stir in the spring onion and season well with salt and freshly ground black pepper. Serve with barbecued sausages, or beef or chicken stews and casseroles.

Fish in ginger broth

SERVES 4

1 tablespoon oil
8 spring onions (scallions), sliced diagonally
2 tablespoons finely chopped ginger
4 tablespoons fish sauce
4 tablespoons grated palm sugar (jaggery)
    or soft brown sugar
4 x 200 g (7 oz) salmon or ocean trout fillets
2 tablespoons lime juice
200 g (7 oz/1 1/3 cups) fresh peas, about
    400 g (14 oz) unpodded
coriander (cilantro) leaves, to garnish

Heat the oil in a large saucepan, add the spring onions and ginger and cook over low heat for 2 minutes, stirring occasionally. Add the fish sauce, sugar and 1.5 litres (52 fl oz/ 6 cups) water and bring to the boil. Reduce the heat, add the fish and poach gently for 3–4 minutes, or until the fish is just cooked through and flakes easily when tested with a fork. Lift the fish out with a slotted spoon and put in a warm shallow bowl. Cover with foil to keep warm.

Bring the liquid in the saucepan to the boil, then reduce the heat and simmer until reduced by half. Add the lime juice.

Cook the peas in a separate saucepan of boiling water for 5 minutes, or until tender, then drain well. To serve, put the fish in bowls, ladle the broth over the top, then divide the peas among the bowls. Garnish the soup with coriander leaves.

brassicas & greens

artichoke

Globe artichokes are the unopened flowers and stems of an edible thistle (if left to bloom, the artichoke transforms into a magnificent purple blossom). Artichokes are best showcased by the simplest of cooking methods — serve simply boiled, with mayonnaise or vinaigrette for dipping their tender inner leaves — but they can also be stuffed and braised or baked, grilled and fried. Artichokes are a spring vegetable, so serve them with other spring produce such as asparagus, broad (fava) beans, peas, and lamb or veal.

Buying and storing

- Artichokes are in season in winter and spring.

- Choose those with green, tightly overlapping leaves that feel heavy for their size. The best test for freshness is to rub the leaves together or to pull one leaf away and to listen for the characteristic 'squeak'.

- Don't buy artichokes if their leaves appear to be opening, as these will be dry. Artichokes start deteriorating from the moment they are picked so freshness is important.

- Use on the day of purchase or store in the fridge in a sealed plastic bag for up to 2 days.

Preparation

Much of a mature artichoke is inedible. The tough outer leaves are removed, the prickly tops are cut off and discarded, as is the hairy 'choke' inside each artichoke. The tender 'heart' of the artichoke is the most prized part, and the tender part of the leaves around it are also eaten.

Preparing an artichoke is fairly simple; it just takes a little time. Snap off the stem, leaving 4–6 cm (1½–2½ inches), and remove any tough fibres. Snap off the tough outer leaves until you reach the paler, tender ones. Cut off about 4 cm (1½ inches) across the top of the artichoke. The stem can then be peeled, and the hairy choke scooped out from the top of the bulb with a teaspoon and discarded. As you go, place the prepared artichokes in water with a squeeze of lemon juice to prevent discolouration.

Cook artichokes in a saucepan of boiling salted water for 20–25 minutes, or until tender through the thickest part (test with a skewer). Cover the artichokes with a plate to keep them submerged. Always use a stainless-steel pan to cook them in, as aluminium can impart a metallic flavour and will discolour the artichokes. Drain, then serve whole with butter or a sauce.

Despite its name, the **Jerusalem artichoke** is neither from Jerusalem nor is it an artichoke. Jerusalem artichokes are a winter root, related to the sunflower, they have creamy brown skin and actually look a bit like fresh ginger. Thinly slice and add raw to salads, boil or roast like potatoes or use to make wonderful velvety soups and mashes. When cut, drop into acidulated water to stop it going brown. Jerusalem artichokes have a reputation for causing flatulence, which can be countered with a pinch of asafoetida spice.

Artichokes vinaigrette

SERVES 4 AS A STARTER

juice of 1 lemon
4 globe artichokes
100 ml (3 1/2 fl oz) olive oil
2 spring onions (scallions), finely chopped
2 tablespoons white wine
2 tablespoons white wine vinegar
1/4 teaspoon dijon mustard
a pinch of sugar
1 tablespoon finely chopped parsley

Bring a large saucepan of salted water to the boil and add the lemon juice. Working with one artichoke at a time, snap off the stems, pulling off any strings, and trim the bases flat. Remove the tough outer leaves. Using kitchen scissors, trim the hard points from the outer leaves, then use a sharp knife to trim the top of the artichoke. Put the artichokes in the water and put a small plate on top of them to keep them submerged. Cook at a simmer for 20–25 minutes, or until a leaf pulls away easily. (The base will be tender when pierced with a skewer.) Cool under cold running water, then drain upside down on a tray.

To make the vinaigrette, heat 1 tablespoon of the oil in a small saucepan, add the spring onion and cook over low heat for 2 minutes. Leave to cool a little, then add the white wine, vinegar, mustard and sugar and gradually whisk in the remaining oil. Season with sea salt and freshly ground black pepper and stir in half the parsley.

Put an artichoke on each plate and gently prise it open a little. Spoon the vinaigrette over the top, allowing it to drizzle into the artichoke and around the plate. Pour the remaining vinaigrette into a small bowl for dipping the leaves into. Sprinkle each artichoke with a little parsley.

Eat the leaves one by one, dipping them in the vinaigrette and scraping the flesh off the leaves between your teeth. When you reach the middle, pull off any really small leaves and then use a teaspoon to remove the furry choke. Once you've got rid of the choke, you can then eat the tender base or 'heart' of the artichoke.

Artichoke, prosciutto and rocket salad

SERVES 4

juice of 1 lemon
4 globe artichokes
2 eggs
3 tablespoons fresh breadcrumbs
3 tablespoons grated parmesan cheese
olive oil, for pan-frying
8 prosciutto slices
3 teaspoons white wine vinegar
1 garlic clove, crushed
150 g (5 1/2 oz) rocket (arugula), stalks
    trimmed
shaved parmesan, to serve

Add the lemon juice to a large bowl of water. Working with one artichoke at a time, trim the stem to 5 cm (2 inches) long and remove the tough outer leaves. Peel the stem using a vegetable peeler. Using kitchen scissors, trim the hard points from the outer leaves, then use a sharp knife to trim across the top. Gently open out the leaves in the centre of the artichoke and use a teaspoon to scrape out the hairy choke. Drop each artichoke into the acidulated water to stop it browning.

Bring a large saucepan of salted water to the boil. Add the artichokes to the boiling water and cook for 2 minutes. Remove using tongs and turn upside down to drain. When cool enough to handle, cut the artichokes into quarters and set aside.

Whisk the eggs in a bowl. In another bowl, combine the breadcrumbs and parmesan, then season with sea salt and freshly ground black pepper. Dip each artichoke quarter into the egg, allowing the excess to drain off, then roll in the breadcrumb mixture to coat.

Fill a heavy-based frying pan with olive oil to a depth of 2 cm (3/4 inch) and heat over medium–high heat. Add the artichokes in batches and cook for 2–3 minutes, or until golden, turning once. Remove with a slotted spoon and drain on paper towels.

Heat another tablespoon of olive oil in a non-stick frying pan over medium–high heat. Add the prosciutto to the pan in two batches and cook for 2 minutes, or until crisp and golden. Remove the prosciutto from the pan, reserving the oil.

Pour the reserved oil into a small bowl. Add the vinegar and garlic, season lightly and whisk to make a dressing. Put the rocket in a bowl, add half the dressing and toss well.

Divide the rocket, artichoke and prosciutto among serving plates. Drizzle with the remaining dressing, scatter with shaved parmesan and sprinkle with a little sea salt.

Artichokes contain cynarin, which makes everything you eat or drink with, or after, them taste sweet. This makes artichokes difficult to match with wine.

- Most Asian greens are available year-round, however the best season for most varieties is during the cooler months of mid-autumn to mid-spring.

- Asian greens should have dark green leaves with no signs of limpness, wilting or tears. Their bases should be dry and firm.

- If possible, use Asian greens on the day of purchase as they don't store well because of their high water content. Otherwise, store in a perforated plastic bag in the crisper of the fridge for 1–2 days.

asian greens

Versatile, nutritious and easy to cook, Asian greens suit stir–frying, steaming, poaching or adding to braises, curries and soups. For a quick, healthy meal, you need little more than a bunch of Asian greens, some seafood, meat or tofu and steamed rice or boiled noodles. Although Asian greens are eaten by more than half the world's population every day, they have only fairly recently become familiar to Western countries, as migration and overseas travel have inspired an appreciation of Asian cuisines, cooking methods and ingredients.

Varieties

Baby bok choy Also known as Shanghai bok choy, this is a small variety of bok choy with upright spoon-shaped green leaves and crunchy pale green stems. The succulent leaves have a mild flavour.

Baby bok choy are at their peak during autumn and spring — steam or braise them whole, then drizzle with oyster sauce, or halve lengthways and stir-fry with ginger, and finish with a little chicken stock and soy sauce. Serve as a side dish to an Asian meal.

Bok choy Also known as pak choy, these have a rounded base, dark green leaves and long, crisp thick stems. Separate the stems and leaves before cooking, as the stems take a little longer to cook, then cook quickly to retain their colour and crunchy texture. Use in soups and stir-fries.

Choy sum Also called Chinese flowering cabbage because of its yellow (or sometimes purple) flowers, choy sum has mild mustard-flavoured leaves and long thin stems that maintain a crunch when properly (that is, briefly) cooked. Use for stir-fries, steaming or poaching.

Gai larn Also called Chinese broccoli, gai larn has dark green leaves, delicate white flowers, and broccoli-like stems that should be peeled and halved lengthways before cooking (young stems are crisp and mild and don't need peeling). Steam whole and serve with oyster sauce or cut up the leaves and stems and add to soups and stir-fries.

Wom bok Variously known as Chinese cabbage, Peking or napa cabbage, wom bok belongs to a different species of brassicas to the European cabbage varieties. Shaped like a cylindrical barrel, wom bok has tightly packed layers of crinkled green and white leaves that are softer than those of European cabbage. Its mild flavour is excellent shredded and eaten raw in salads, lightly steamed, stir-fried or added to hotpots and clear Asian soups.

Water spinach An edible aquatic vine that grows in moist garden beds or suspended in water, water spinach is also known as water convolvus, ong choi, kang kung or pak bung, depending on where you are. It has dark green, pointy, elongated, narrow leaves and hollow, long stems. To cook water spinach, first wash it well, then cut the leaves and stems into 5 cm (2 inch) lengths and add to Thai-style curries, Asian soups or stir-fries.

Gai choy Also called mustard cabbage because of its distinctive mustardy flavour, gai choy has thick, bulbous pale green stems with crinkled, broad light green leaves. Traditionally it is braised, added to soups or pickled, but can also be stir-fried. Some cooks prefer to blanch it first to temper its strong flavour.

Preparation

When cooking varieties of Asian greens with thick stems, it is best to separate the leaves from the stems and cook the stems first, as they take longer to cook. Asian greens should be trimmed before cooking, then soaked in cold water to remove any dirt.

Asian greens suit stir-frying or steaming but be careful not to overcook them: the leaves should be just wilted and the stems should still have a bit of crunch — this may only take a minute or two in the wok. Baby bok choy can be steamed whole, which will take about 6–10 minutes.

Stir-fried tofu with choy sum

SERVES 4

3 tablespoons vegetable oil
2 tablespoons lime juice
1$\frac{1}{2}$ tablespoons fish sauce
1 teaspoon sambal oelek
$\frac{1}{2}$ teaspoon soft brown sugar
200 g (7 oz) smoked tofu
400 g (14 oz/1 bunch) choy sum (Chinese
    flowering cabbage), trimmed
150 g (5$\frac{1}{2}$ oz/4 cups) torn English spinach
2 teaspoons sesame seeds, toasted
1 small handful coriander (cilantro)

To make the dressing, put 2 tablespoons of the vegetable oil, the lime juice, fish sauce, sambal oelek and sugar in a bowl and whisk well to dissolve the sugar.

Cut the smoked tofu into 2 cm ($\frac{3}{4}$ inch) cubes. Trim the choy sum and cut into 8 cm (3$\frac{1}{4}$ inch) lengths.

Heat the remaining 1 tablespoon of oil in a large wok over medium heat and gently stir-fry the tofu for 2–3 minutes, or until golden brown. Add half the dressing and toss to coat. Remove from the wok and set aside.

Add the choy sum to the wok and stir-fry for 1 minute. Add the spinach leaves and stir-fry for 1 minute. Return the tofu to the wok, add the sesame seeds and the remaining dressing and toss lightly. Serve garnished with the coriander leaves.

Gai larn in oyster sauce

SERVES 6

1 kg (2 lb 4 oz) gai larn (Chinese broccoli)
125 ml (4 fl oz/$\frac{1}{2}$ cup) chicken stock
3 tablespoons oyster sauce
2 tablespoons light soy sauce
1 tablespoon Chinese rice wine
1 teaspoon sugar
1 teaspoon sesame oil
2 teaspoons cornflour (cornstarch)
1$\frac{1}{2}$ tablespoons oil
2 spring onions (scallions), finely chopped
1$\frac{1}{2}$ tablespoons grated ginger
3 garlic cloves, finely chopped

Wash the gai larn well. Discard any tough-looking stems, then diagonally cut into 2 cm ($\frac{3}{4}$ inch) pieces through the stem and the leaf. Blanch in a saucepan of boiling water for 2 minutes, or until the stems and leaves are just tender, then refresh in cold water and dry thoroughly.

Combine the stock, oyster sauce, soy sauce, rice wine, sugar and sesame oil in a bowl, stirring to dissolve the sugar. Blend a little of the mixture into the cornflour, then pour the cornflour into the bowl, stirring to combine. Set aside.

Heat a wok over high heat, add the oil and heat until very hot. Add the spring onion, ginger and garlic and stir-fry for 10 seconds, or until fragrant, then add the gai larn and cook until heated through. Pour the sauce mixture into the wok, stirring until the sauce has thickened, and toss to coat the gai larn.

Asian greens work best when matched with other Asian flavours or ingredients, especially rice, noodles, chilli, garlic, ginger, black bean sauce, hoisin sauce, oyster sauce, fish sauce, soy sauce, pork, chicken, duck, seafood, rice wine, tofu, shiitake mushrooms and sesame oil.

Gai larn with ginger, lime and peanuts

SERVES 4

40 g (1½ oz) tamarind pulp
3 tablespoons boiling water
1 tablespoon peanut oil
600 g (1 lb 5 oz/1 large bunch) gai larn
   (Chinese broccoli), trimmed and halved
   widthways
1 small red chilli, seeded and finely chopped
2 garlic cloves, finely chopped
3 teaspoons finely grated ginger
1 tablespoon sugar
1 tablespoon lime juice
1 teaspoon sesame oil
1 tablespoon finely chopped roasted
   unsalted peanuts

Put the tamarind in a bowl and pour in the boiling water. Set aside to soak for 5 minutes, then strain. Discard the solids.

Heat a non-stick frying pan over high heat, add the oil and swirl to coat. Add the gai larn and stir-fry for 2–3 minutes, or until wilted. Add the chilli, garlic and ginger and cook for another minute, then add the sugar, lime juice and 1 tablespoon of the tamarind liquid and simmer for 1 minute. Transfer to a plate and drizzle with the sesame oil. Scatter with peanuts to serve.

Beef and bok choy

SERVES 4

600 g (1 lb 5 oz/1 large bunch) bok choy
   (pak choy)
2 tablespoons oil
2 garlic cloves, crushed
250 g (9 oz) rump steak, thinly sliced
2 tablespoons soy sauce
1 tablespoon sweet sherry
2 tablespoons chopped basil
2 teaspoons sesame oil
¼ red capsicum (pepper), sliced, to serve

Wash the bok choy and drain, then cut the leaves into strips. Heat 1 tablespoon of the oil in a wok over medium heat, add the garlic and stir-fry for 30 seconds.

Add the remaining oil to the wok. When the oil is hot, add the beef in small batches and stir-fry for 3 minutes over high heat until the meat is just browned but not cooked through. Remove the meat from the wok.

Add the bok choy and stir-fry for 30 seconds, or until it is just wilted. Return the beef to the wok along with the soy sauce and sherry. Stir-fry for 2–3 minutes, or until the beef is tender. Add the basil and sesame oil and toss well. Serve immediately garnished with strips of red capsicum.

asparagus

A member of the lily family, these delicately flavoured shoots are best prepared simply — boiled, steamed or pan-fried, then served as an accompaniment, or cooled and used in salads or other dishes. Hot, cooked asparagus is best served simply tossed in extra virgin olive oil or melted butter, or served with a sauce such as hollandaise.

Varieties

There are over 300 species of asparagus, including some wild varieties. Green asparagus is the most common but there are also purple and white varieties. White asparagus is actually the same type as green, except for the way it is grown. The white variety is grown in deep soil trenches or tunnels, hidden from sunlight, thus interrupting the greening process of photosynthesis — this is called 'blanching'. Purple asparagus, which turns green when cooked, tastes fairly similar to green asparagus.

Buying and storing

- Resist all temptation to eat out-of-season asparagus: these delicately flavoured shoots are best eaten in spring, just after harvesting.

- Asparagus tips should be tight — don't buy them if the spears look a little floppy. The stem end should not be wrinkled or dried out.

- Choose spears roughly the same size so they will cook at the same time.

- Buy asparagus the day you plan to eat it, or store for 2 days in a plastic bag in the fridge.

Preparation

Snap off the woody end of the asparagus spear at its natural breaking point, and peel any thick, woody stems. White asparagus spears always need peeling, as their skin is tough.

Cook asparagus in boiling salted water for about 5 minutes, or until tender when pierced with the tip of a knife. Use a wide saucepan or a frying pan for cooking so the spears can easily fit laying flat. A well-cooked spear of asparagus should be tender and bend when you pick it up with a fork; it should not droop over. Drain well. Asparagus can also be steamed or pan-fried for 5–7 minutes, until tender.

Andalucian asparagus

SERVES 4 AS A STARTER OR SIDE DISH

500 g (1 lb 2 oz/about 3 bunches)
    asparagus spears
1 thick slice country-style bread
3 tablespoons extra virgin olive oil
2–3 garlic cloves
12 blanched almonds
1 teaspoon paprika
1 teaspoon ground cumin
1 tablespoon red wine vinegar or sherry
    vinegar

Snap off the woody ends from the asparagus. Remove the crusts from the bread and cut the bread into cubes.

Heat the olive oil in a frying pan over medium heat. Add the bread, garlic and almonds and sauté over medium heat for 2–3 minutes, or until golden.

Using a slotted spoon, transfer the bread, garlic and almonds to a food processor. Add the paprika, cumin, vinegar, 1 tablespoon water and some sea salt and freshly ground black pepper. Process until the mixture is finely chopped.

Return the frying pan to the heat and add the asparagus, adding a little extra olive oil if necessary. Cook over medium heat for 3–5 minutes, or until just cooked, turning often. Transfer to a serving plate.

Add the almond mixture to the frying pan with 200 ml (7 fl oz) water. Simmer for 2–3 minutes, or until the liquid has thickened slightly. Spoon over the asparagus and serve.

Warm asparagus with creamy orange-pepper dressing

SERVES 4 AS A SIDE DISH

2 tablespoons extra virgin olive oil
2 tablespoons orange juice
1 teaspoon finely grated orange zest
3 tablespoons crème fraîche
20 asparagus spears, trimmed

Preheat a barbecue grill or chargrill pan to medium.

In a small bowl, combine the olive oil and orange juice, and season well with sea salt and freshly ground black pepper. Pour half the mixture into a shallow dish for coating the asparagus.

Combine the orange zest and crème fraîche with the remaining juice mixture, then season to taste with sea salt and freshly ground black pepper and set aside.

Put the asparagus in the oil mixture in the dish and toss to coat. Remove and cook the asparagus on the barbecue, turning often, for 4–5 minutes, or until tender and lightly charred. Transfer to a serving platter, pour the dressing over and serve immediately.

Asparagus and mint frittata

SERVES 4

6 eggs
35 g (1¼ oz/⅓ cup) grated pecorino or
    parmesan cheese
1 handful mint, finely shredded
200 g (7 oz) baby asparagus spears
    (about 16)
2 tablespoons extra virgin olive oil

Break the eggs into a large bowl, beat well, then stir in the cheese and mint. Set aside. Snap off and discard the woody ends of the asparagus, then cut on the diagonal into 5 cm (2 inch) pieces.

Heat the olive oil in a frying pan that has a heatproof handle. Add the asparagus and cook for 5 minutes, or until the asparagus is tender. Season, then reduce the heat to low. Pour the egg mixture over the asparagus and cook for 8–10 minutes. Use a spatula to pull the side of the frittata away from the side of the pan and tip the pan slightly so the uncooked egg runs underneath the frittata.

When the mixture is nearly set but still slightly runny on top, place the pan under a low grill (broiler) for 1–2 minutes, until the top is set. Serve warm or at room temperature.

Tagliatelle with asparagus and herbs

SERVES 4–6

500 g (1 lb 2 oz) tagliatelle pasta
40 g (1½ oz) butter
200 g (7 oz) asparagus spears (about 16),
    trimmed and cut into short pieces
1 tablespoon chopped parsley
1 tablespoon chopped basil
310 ml (10¾ fl oz/1¼ cups) cream
50 g (1¾ oz/½ cup) finely grated parmesan
    cheese
shaved parmesan cheese, to serve

Cook the pasta in a large saucepan of boiling salted water until al dente. Drain and return to the pan.

Heat the butter in a frying pan, add the asparagus and stir over medium heat for 2 minutes, or until just tender, then add the parsley, basil and cream, and season with salt and freshly ground black pepper. Cook for a further 2 minutes.

Add the parmesan to the pan and stir well. When thoroughly mixed, add the sauce to the warm pasta and toss gently to distribute the ingredients evenly. Serve in warmed bowls with shavings of parmesan cheese.

broccoli

Broccoli was first cultivated by the Italians and its name is the Italian word for 'little arms'. Broccoli is a member of the brassica family, related to cauliflower and cabbage, and was originally a form of wild cabbage that eventually evolved to develop buds rather than leaves.

Varieties

The most common broccoli is the calabrese, with tightly clustered emerald green heads. There is also a purple-tinged variety and a lime green one, as well as Chinese broccoli, gai larn. A more recent addition is broccolini, a cross between broccoli and gai larn, popular for its tender long stalks, deliciously subtle flavour and loose, small buds.

Buying and storing

- Broccoli is available year-round, however its peak season is from winter to spring.

- Look for broccoli with firm, tightly closed green heads, with no signs of yellowing. The stalks should be firm and not too thick or they will be woody.

- Broccolini should have long, firm stems and a smallish, compact head. The florets should be blue-green — a few yellow flowers per bunch are acceptable but more than this indicates overmaturity.

- Store in a perforated or open plastic bag in the crisper of the fridge for up to 4 days.

Preparation

Use a small knife to separate the florets from the stalks, cutting the florets into even-sized pieces. Soak the florets in lightly salted water to remove any bugs or caterpillars. The stalks are usually thrown away, but these can be peeled, sliced and steamed (they will take a few minutes longer to cook). Cook in boiling salted water for 3–4 minutes, until tender but still bright green. Drain well, as the florets hold lots of water.

To prepare broccolini, simply wash it and trim off the base of the stem. It can then be steamed or boiled whole for 3–4 minutes, or chopped and stir-fried.

Broccoli and almond stir-fry

SERVES 4

1 teaspoon coriander seeds
3 tablespoons olive oil
2 tablespoons slivered almonds
1 garlic clove, crushed
1 teaspoon finely shredded ginger
500 g (1 lb 2 oz) broccoli, cut into florets
2 tablespoons red wine vinegar
1 tablespoon soy sauce
2 teaspoons sesame oil
1 teaspoon sesame seeds, toasted

Lightly crush the coriander seeds using a mortar and pestle. Heat the olive oil in a wok or large heavy-based frying pan. Add the coriander seeds and almonds and stir quickly over medium heat for 1 minute, or until the almonds are golden.

Add the garlic, ginger and broccoli to the wok. Stir-fry over high heat for 2 minutes, then remove the wok from the heat. Pour the combined vinegar, soy sauce and sesame oil into the wok. Toss until the broccoli is well coated, then serve immediately, sprinkled with toasted sesame seeds.

Broccoli and pine nut soup

SERVES 6 AS A STARTER

30 g (1 oz) butter
1 onion, finely chopped
1.5 litres (52 fl oz/6 cups) chicken stock
750 g (1 lb 10 oz) broccoli, trimmed
4 tablespoons pine nuts, plus extra to serve
extra virgin olive oil, to serve
toasted focaccia, to serve

Melt the butter in a large saucepan. Add the onion and sauté over medium heat for 5 minutes, or until softened but not browned. Add the stock and bring to the boil.

Cut the florets from the broccoli and set aside. Chop the broccoli stalks and add them to the pan, then reduce the heat, cover and simmer for 15 minutes. Add the florets and simmer, uncovered, for 10 minutes, or until the florets are tender. Remove from the heat and allow to cool completely.

Add the pine nuts to the soup. Transfer to a blender or food processor, in batches if necessary, and blend until smooth. Season to taste with sea salt and freshly ground black pepper, then gently reheat. Serve sprinkled with extra pine nuts and drizzled with extra virgin olive oil. Serve with toasted focaccia.

Broccoli is one of the healthiest of vegetables, containing excellent doses of folate and high levels of vitamin A. It is also rich in carotene, and high intakes may provide protection against cancer and heart disease. A single cup of broccoli boasts as much calcium as a cup of milk.

Broccoli and ricotta soufflé

SERVES 4

60 g (2 1/4 oz/1 cup) small broccoli florets
2 tablespoons olive oil
40 g (1 1/2 oz) butter
1 onion, finely chopped
1 garlic clove, crushed
400 g (14 oz/scant 1 2/3 cups) ricotta cheese
50 g (1 3/4 oz/1/2 cup) grated parmesan cheese
5 egg yolks, lightly beaten
a pinch of ground nutmeg
a pinch of cayenne pepper
5 egg whites
a pinch of cream of tartar
3 tablespoons dry breadcrumbs

Preheat the oven to 190°C (375°F/Gas 5).

Cook the broccoli florets in boiling salted water for 4 minutes, then drain well and roughly chop.

Heat the olive oil and butter in a frying pan. Add the onion and garlic and sauté over medium heat for 5 minutes, or until the onion has softened. Transfer to a large bowl and add the broccoli, ricotta, parmesan, egg yolks, nutmeg and cayenne pepper. Season with sea salt and freshly ground black pepper. Mix well.

In a clean, dry bowl, whisk the egg whites with the cream of tartar and a pinch of salt until stiff peaks form. Stir one-third of the beaten egg white into the broccoli mixture to loosen, then gently fold in the remaining egg white.

Grease a 1 litre (35 fl oz/4 cup) soufflé dish. Sprinkle with the breadcrumbs, turn the dish to coat, then shake out the excess. Spoon the broccoli mixture into the dish and bake for 35–40 minutes, or until puffed and golden brown. Serve immediately.

note This soufflé is based on ricotta cheese so it won't rise as much as a conventional soufflé would.

brussels *sprout*

Love them or loathe them, when it comes to brussels sprouts, the culinary world is clearly divided. Like their cabbage cousins, overcooked sprouts give off a sulphurous smell (hence, their bad reputation) but if handled correctly they can be a sweet revelation. Brussels sprouts were first cultivated in Flanders, near Brussels.

Buying and storing

- The best time to buy brussels sprouts is during winter and spring.

- Choose firm, bright green brussels sprouts with no splitting, browning or yellowing on the outer leaves.

- Buy smaller ones as they are more tender and tastier.

- Loosely wrap brussels sprouts in paper towels, then place in a perforated plastic bag and store in the crisper of the fridge for up to 3 days.

Preparation

Wash the sprouts well before cooking, or if they have come directly from the garden, soak them in cold water in case they contain a few worms. Remove any discoloured outer leaves and score a cross into their base to promote even cooking.

Boiling Cook the sprouts whole, uncovered, in plenty of boiling salted water for 6–8 minutes, or until just tender, then drain well. Cooking with the lid off helps them to stay green.

Steaming This is a good way to par-cook brussels sprouts if you then want to sauté them, bake them in a gratin or want to finish them in a creamy sauce. Steam them halved or whole for 8–12 minutes, or until just tender.

Frying For a quick side dish, fry 400 g (14 oz) shredded brussels sprouts in a little olive oil until tender. Add 4 finely chopped bacon slices and fry together until crisp. To serve, season with freshly ground black pepper and sprinkle a few chopped almonds over the top.

The sprouts can also be sliced and stir-fried with flavours such as garlic, soy sauce and sesame seeds. They also pair well with ingredients such as bacon, chives, shallots, cream, garlic, horseradish, mustard, nutmeg and roast pork.

Bubble and squeak

SERVES 4

750 g (1 lb 10 oz) floury potatoes
125 ml (4 fl oz/^{1}/$_2$ cup) milk
90 g (3^1/$_4$ oz) butter
450 g (1 lb) mixed green vegetables, such
    as English spinach, brussels sprouts,
    cabbage and leek, thinly sliced

Brussels sprouts with pancetta

SERVES 4 AS A SIDE DISH

100 g (3^1/$_2$ oz) pancetta, thinly sliced
4 French shallots
20 g (^{3}/$_4$ oz) butter
1 tablespoon olive oil
1 garlic clove, crushed
500 g (1 lb 2 oz) brussels sprouts,
    trimmed and thickly sliced

Preheat the grill (broiler) to hot. Spread the pancetta on a baking tray or on a grill rack lined with foil and put the tray 8–10 cm (3^1/$_4$–4 inches) under the heat. Grill (broil) for 1 minute, or until crisp. Set aside to cool.

Peel the shallots and slice them into thick rings. Heat the butter and oil in a large frying pan over medium heat. Add the shallots and garlic and cook for 3–4 minutes, or until just starting to brown. Add the brussels sprouts and season with freshly ground black pepper. Cook, stirring often, for 4–5 minutes, or until lightly golden and crisp. Turn off the heat, cover and set aside for 5 minutes.

Break the pancetta into large shards. Add to the brussels sprouts and toss lightly; some of the pancetta will break up into smaller pieces. Serve immediately.

Cut the potatoes into even-sized pieces and put them in a saucepan of cold water. Bring to the boil, then lower the heat and simmer until tender — do not boil or the potatoes may break up and absorb water before they cook. Drain thoroughly.

Heat the milk in the pan. Return the potatoes to the pan with half the butter, then mash until the potato is smooth and creamy.

Melt half of the remaining butter in a large heavy-based frying pan with a heatproof handle, and cook the green vegetables until they are tender and cooked through. Add them to the potato and mix together. Season with salt and freshly ground black pepper.

Melt the remaining butter in the frying pan and spoon in the potato mixture, smoothing off the top. Cook until the bottom is browned and crispy. Remove the pan from the heat and place it under a hot grill (broiler) and cook until the top is browned and golden. If you prefer, you can turn the bubble and squeak over in the pan and cook it on the other side, but grilling is easier. Serve as a meal by itself or as an accompaniment to poached eggs.

cabbage

Cabbage leaves wrapped around various permutations of rice, minced meat, herbs and spices are a time-honoured tradition from Eastern Europe to the Middle East, while the fermented cabbage staple, sauerkraut, is used in Central European cooking.

Varieties

Cavolo nero This dark blackish cabbage with thick, long, curly leaves has a rich, rather bitter taste. Briefly sauté cavolo nero in oil with garlic and chilli, or braise in the oven. Serve drizzled with good-quality olive oil and a generous grinding of black pepper as an accompaniment to meat and chicken dishes. Cavolo nero is an essential ingredient in Tuscan cooking in soups and stews, notably in the sustaining vegetable soup called ribollita.

Green or white These are the most common types of cabbages; their rounded heads are comprised of tightly packed leaves, thick stems and cores. Large green cabbages are suitable for stuffing. Baby or dwarf green cabbages are more delicately flavoured than their larger cousins and are suited to minimal cooking

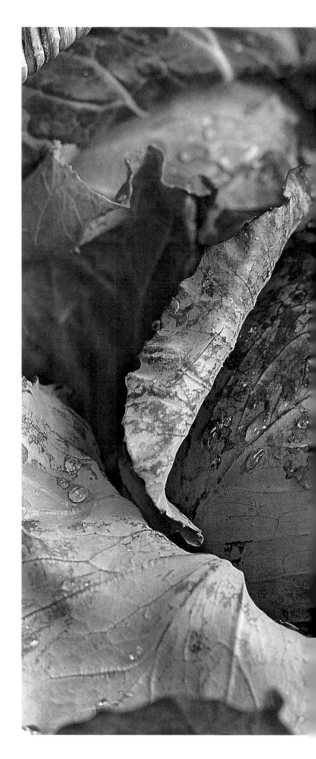

methods such as stir-frying, blanching or steaming. Sometimes called Dutch cabbage, the white cabbage's smooth, crisp and crunchy leaves are ideal for using raw in salads. This is the preferred cabbage for sauerkraut.

Red A brilliant purplish red cabbage with crisp leaves that have a mellow taste, red cabbage is a vibrant addition to winter salads and stir-fries. Cook it quickly to maintain its colour. If used in slow-cooked soups and stews, add an acidic ingredient such as lemons, apples, wine or vinegar to preserve its colour; otherwise, it turns a greyish blue.

Bake thickly sliced red cabbage in a casserole dish with red wine vinegar, onions and dried prunes. Serve with pork or beef casseroles. Red cabbage is used to make a vinegary pickle popularly served with cold meat and game.

Savoy A dark green, loose-headed cabbage with attractive large leaves and a mild flavour. Blanched savoy cabbage leaves make ideal 'wrappers' for stuffing mixtures such as rice and minced pork or veal. Sauté chopped leaves with porcini mushrooms and sage and stir through pasta or add to risotto with creamy fontina cheese and pancetta.

The Romans ate **cabbage** to fend off melancholy. Whatever your state of mind, it goes well with apples, anchovies, bacon, cheese, cream, sour cream, caraway seeds, cider vinegar, chestnuts, nutmeg, bacon, pancetta, pork, veal and chicken.

Cabbage contains sulphur compounds, which are released when the vegetable is boiled; the longer the boiling, the greater the amount of compounds released into the air. To keep things sweet in the kitchen, keep boiling times to a minimum or cook the cabbage using other cooking methods, such as stir-frying, steaming or braising.

Buying and storing

- Tight-leaved green and white cabbages are available year-round but are best during winter; savoy is best during spring; and autumn is the best time to buy red cabbage.

- Choose firm cabbages that are heavy for their size with no yellow or brown patches.

- For tight-leaf cabbages such as the green and red varieties, choose ones with tightly closed outer leaves enclosing their heads. Store in a plastic bag in the crisper of the fridge for up to 1 week.

- For loose-leaf varieties such as cavolo nero and savoy, choose ones with crisp, strongly coloured leaves. Store in a plastic bag in the fridge for 2–3 days.

Preparation

Remove the tough outer leaves and any discoloured or wilted ones. Cut the cabbage in half lengthways and cut out the hard central core. If using whole leaves for stuffing, cut them away at the base, easing them carefully off the head with your fingers. They are very prone to tearing, so if using them as a wrap for stuffings, prepare more than the recipe states.

To precook leaves for stuffing, cook them a few at a time in a large saucepan of boiling water for 3–5 minutes, or until just tender, then drain and pat well to dry.

Chestnut, pancetta and cabbage soup

SERVES 4

100 g (3 1/2 oz) savoy cabbage, roughly
    chopped
2 tablespoons olive oil
1 large onion, finely chopped
185 g (6 1/2 oz) pancetta, diced
3 garlic cloves, crushed
2 tablespoons chopped rosemary
300 g (10 1/2 oz/2 cups) cooked peeled
    chestnuts
150 ml (5 fl oz) red wine
extra virgin olive oil, to serve

Bring 1.5 litres (52 fl oz/6 cups) salted water to the boil, add the cabbage and cook over high heat for 10 minutes. Drain, reserving the water. Cool the cabbage slightly, then finely chop.

Heat the olive oil in a large saucepan, add the onion and pancetta and cook over medium–high heat for 4–5 minutes, or until the onion is soft and the pancetta is lightly browned. Add the garlic and rosemary and cook for a few more minutes.

Using your hands, break up the chestnuts and add to the pan along with the cabbage. Stir to combine, season with salt and freshly ground black pepper, then add the wine. Bring to the boil and cook for 2 minutes. Add the cabbage water, then reduce the heat and simmer for 15 minutes. Remove half the soup mixture from the saucepan and allow to cool slightly before blending to a purée in a food processor. Return the purée to the pan over heat and stir to combine and heat through. Serve with a drizzle of extra virgin olive oil.

Classic coleslaw

SERVES 4

1/2 small green cabbage
1/4 small red cabbage
2 carrots, coarsely grated
4 radishes, coarsely grated
1/2 red capsicum (pepper), chopped
3 spring onions (scallions), sliced
3 tablespoons chopped parsley
175 g (6 oz/2/3 cup) ready-made egg
    mayonnaise

Remove the hard core from the cabbages and shred the leaves with a sharp knife. Put the cabbage in a large bowl and add the carrot, radish, capsicum, spring onion and parsley. Toss to combine, then refrigerate until ready to serve.

Just before serving, add the mayonnaise and season to taste with sea salt and freshly ground black pepper. Toss together until well combined, then serve.

Pork chops with braised red cabbage

SERVES 4

Pork with apple and cabbage is a classic combination, the cabbage providing a sweet foil for the richness of the pork. The cabbage is braised slowly to bring out its sweetness.

braised red cabbage
1 tablespoon oil
1 onion, finely chopped
1 garlic clove, crushed
1 small red cabbage, shredded
1 dessert (eating) apple, peeled, cored
    and thinly sliced
4 tablespoons red wine
1 tablespoon red wine vinegar
$1/4$ teaspoon ground cloves
1 tablespoon chopped sage

1 tablespoon oil
4 x 200 g (7 oz) pork chops, trimmed
4 tablespoons white wine
375 ml (13 fl oz/$1^1/2$ cups) chicken stock
3 tablespoons thick (double/heavy) cream
$1^1/2$ tablespoons dijon mustard
4 sage leaves

To make the braised cabbage, heat the oil in a large saucepan, add the onion and garlic and cook, stirring often, over medium heat for 6 minutes, or until the onion is softened but not browned. Add the cabbage, apple, red wine, vinegar, cloves and sage and season with salt and freshly ground black pepper. Cover the pan and cook for 30 minutes over very low heat. Uncover the pan and cook, stirring, for a further 5 minutes, increasing the heat if necessary, to evaporate any liquid.

Meanwhile, heat the oil in a frying pan. Season the pork, then add to the pan and cook over medium–high heat for about 5 minutes, turning once, until the pork is browned on both sides. Add the white wine and stock, cover and simmer for 20 minutes, or until the pork is tender.

Remove the pork from the frying pan and strain the liquid. Return the liquid to the pan, bring to the boil and cook until reduced by two-thirds. Add the cream and mustard and stir over very low heat, without boiling, until the sauce has thickened slightly. Pour over the pork chops and garnish with sage. Serve with the red cabbage.

cauliflower

Like broccoli, the edible part of cauliflower is made up of tightly clustered florets of immature flower buds called the 'curd', surrounded by tight green leaves. These leaves protect the cauliflower from the sun, preventing the development of chlorophyll (which would turn it green), thus maintaining its characteristic milky white colour.

Varieties

Besides the common white cauliflower, there are also green and purple varieties, as well as miniature ones and a broccoflower, which, as its name suggests, is a cross between broccoli and cauliflower. This is light green in colour and milder in taste than either of its 'parents'.

Buying and storing

- Cauliflower is available year-round but is best in winter.

- Buy those with compact, creamy white heads (curds) with no browning. They should smell fresh. Avoid any with black spots, as this indicates spoilage or water damage.

- Remove the outermost leaves and store in a perforated plastic or paper bag in the fridge for up to 5 days.

Preparation

Remove the outer leaves and any tough stalk, then cut into florets. Soak the florets in lightly salted water to get rid of any bugs.

Cauliflower contains a natural chemical that releases an unpleasant sulphurous smell when cooked. To prevent this, cook until just tender — the longer it cooks the stronger the smell becomes. Some cooks add a bay leaf to help prevent this. Use a non-aluminium saucepan as aluminium reacts with cauliflower and can turn it an unsightly brownish hue.

Cauliflower can be steamed or boiled but steaming keeps the florets intact. It can also be roasted, which gives it a nutty sweetness and none of the trademark sulphurous smell. Put the florets in a baking dish, drizzle generously with olive oil and cook in a 180°C (350°F/Gas 4) oven for about 30 minutes, turning the cauliflower often, or until soft and golden.

Cauliflower and pasta bake

SERVES 6

150 g (5^1/2 oz/1^2/3 cups) penne pasta
600 g (1 lb 5 oz) cauliflower, cut into florets
2 tablespoons olive oil
2 red onions, chopped
2 garlic cloves, finely chopped
80 g (2^3/4 oz) butter
4 tablespoons plain (all-purpose) flour
1 litre (35 fl oz/4 cups) milk
200 g (7 oz/2 cups) grated parmesan
   cheese
3 handfuls basil, torn
5 slices day-old bread, crusts removed
50 g (1^3/4 oz) butter, melted

Preheat the oven to 180°C (350°F/Gas 4). Cook the pasta in boiling salted water until al dente, then drain and set aside. Steam the cauliflower for 10 minutes, or until just tender.

Heat the olive oil in a frying pan over medium heat. Fry the onion and garlic for 5 minutes, or until the onion is soft. Combine in a bowl with the cauliflower.

Melt the butter in a large frying pan. Stir in the flour and cook, stirring constantly, for 1 minute, then gradually whisk in the milk. Stir constantly until the mixture boils and thickens. Remove from the heat and add 125 g (4^1/2 oz) of the parmesan and the basil and stir through. Add the cauliflower, pasta and onions and mix thoroughly.

Spoon the cauliflower and pasta mixture into a large ovenproof dish. Cut the bread into large cubes, toss in the melted butter, then scatter over the cauliflower mixture. Sprinkle with the remaining parmesan and bake for 35–40 minutes, or until the top is golden.

Cauliflower fritters

SERVES 4–6 AS A STARTER OR SIDE DISH

55 g (2 oz/^{1}/2 cup) besan (chickpea flour)
   (see note)
2 teaspoons ground cumin
1 teaspoon ground coriander
1 teaspoon ground turmeric
a pinch of cayenne pepper
1 egg
1 egg yolk
600 g (1 lb 5 oz) cauliflower, cut into
   bite-sized florets
oil, for deep-frying

Sift the besan, spices and ^{1}/2 teaspoon sea salt into a bowl. Make a well in the centre. Lightly whisk the egg and egg yolk with 3 tablespoons water, then pour into the well, whisking until the batter is smooth. Set aside for 30 minutes.

Meanwhile, preheat the oven to 120°C (235°F/Gas ^{1}/2).

Fill a large heavy-based saucepan one-third full of oil and heat to 180°C (350°F), or until a cube of bread dropped into the oil browns in 15 seconds. Dip the cauliflower florets into the batter, allowing the excess to drain off, then deep-fry in batches for 3 minutes, or until puffed and browned. Remove with a slotted spoon and drain on paper towels, then place in the oven to keep warm while cooking the remaining fritters. Serve hot, sprinkled with sea salt.

note Besan is available from health food stores and Indian grocers.

radicchio

Radicchio is the Italian word for 'red chicory' and is related to chicory, curly endive and witlof (Belgian endive). Radicchio adds a bold magenta colour and distinctive bitter flavour to dishes such as salads or risottos. Its characteristic bitterness is due to a chemical called intybin, which stimulates the appetite and digestive system.

Varieties

Radicchio is associated with the Veneto region of Italy and the different varieties are named after towns in that region. Radicchio di chioggia, a deep garnet-coloured, round-headed radicchio, and radicchio di castelfranco, a soft, crumpled and speckle-leaved variety, are often used in salads for their decorative, red-tinged green leaves and mildly flavoured pale hearts. Radicchio di verona has elongated, bright red leaves and radicchio di treviso, a more elongated version with thick white ribs and maroon leaves, is best suited to cooking.

Buying and storing

- Radicchio is in season from late autumn to winter.

- Choose radicchio that shows no signs of browning or wilting, with unbroken, crisp and brightly coloured leaves. Inspect the stem end to make sure it looks fresh and is not slimy or browned.

- Store radicchio like lettuces, in perforated plastic bags in the crisper of the fridge for up to 2 days.

Preparation

Remove the tough outer leaves and discard them, then remove the inner leaves and wash them to remove all dirt. Pat dry with paper towels or dry the leaves in a salad spinner.

While radicchio is often used in green salads, its flavour mellows and sweetens when cooked. Try halving treviso radicchio lengthwise, drizzling with olive oil and grilling, then serving with barbecued quail, tuna steaks or lamb. Sauté strips of radicchio with butter, garlic and prosciutto and toss through cooked spaghetti; and thinly sliced radicchio makes an invigorating last-minute addition to a barley soup.

Radicchio goes well with ingredients such as almonds, pasta, rice, parmesan or goat's cheese, hazelnut oil, artichokes, rocket (arugula), watercress, chicken, duck, swordfish and tuna.

Chargrilled radicchio

SERVES 4 AS A SIDE DISH

2 heads radicchio
3 tablespoons olive oil
1 teaspoon balsamic vinegar

Trim the radicchio, discarding the outer leaves. Slice the radicchio into quarters lengthways and rinse well. Drain, then pat dry with paper towels.

Heat a chargrill pan or frying pan over high heat. Lightly sprinkle the radicchio with some of the olive oil and season with sea salt and freshly ground black pepper, then cook for 2–3 minutes, until the under leaves soften and darken. Turn and cook the other side. Transfer to a dish and sprinkle with the remaining oil and vinegar.

Serve hot with grilled meats, or cold as part of an antipasti selection.

Radicchio with figs and ginger vinaigrette

SERVES 4

1 head radicchio
1 baby frisée (curly endive)
3 oranges or mandarins
1/2 small red onion, thinly sliced into rings
8 small green figs, cut into quarters
3 tablespoons extra virgin olive oil
1 teaspoon red wine vinegar
2 tablespoons orange juice
a pinch of ground cinnamon
2 tablespoons very finely chopped glacé ginger, plus 2 teaspoons syrup
1 pomegranate (optional)

Wash the radicchio and frisée leaves thoroughly and drain well. Tear any large leaves into bite-sized pieces and toss in a salad bowl.

Peel and segment the oranges, discarding all the bitter white pith. Add to the salad leaves with the onion and 8 of the fig quarters, reserving the remaining fig quarters.

Whisk the oil, vinegar, orange juice, cinnamon, ginger and ginger syrup in a small bowl. Season with sea salt and freshly ground black pepper. Pour over the salad and toss lightly.

Arrange the reserved figs in pairs over the salad. If you are using the pomegranate, cut it in half, scoop out the seeds and scatter them over the salad before serving.

silverbeet

From the filo pies of Greece, the vegetable frittatas of Italy, and shredded silverbeet stirred into the lentil soups and chickpea casseroles of the Middle East, silverbeet finds itself in myriad dishes. Although sometimes called Swiss chard, silverbeet actually originated in the Mediterranean, not Switzerland, as the name might suggest. While many people cook just the leaves and the tender white ribs, in many parts of Europe the tougher stalks are actually considered the most prized part of the plant — cook them separately, as they usually take longer than the leaves to cook.

Buying and storing

- Although you can find it in markets year-round, the peak season for silverbeet is from mid-autumn to spring.

- Choose silverbeet with shiny, firm leaves that show no signs of wilting or yellowing. The stalks should be unblemished and crisp-looking.

- Although silverbeet grows well under diverse conditions and will even thrive if neglected, it is not so robust once picked. Its leaves are easily bruised and it is best not to keep it for too long — store in a perforated plastic bag in the crisper of the fridge for up to 3 days. If preferred, cut off the long stalks and store them separately.

Preparation

Silverbeet holds lots of dirt in its leafy folds, so needs to be washed well. Dry, then lay several leaves on top of each other, roll up tightly and slice. Cook the leaves, covered, in a very small quantity of boiling water for 5 minutes, or until tender, then transfer to a colander and press down on it with the back of a large spoon to extract as much liquid as possible.

To cook the stalks, trim several centimetres off the ends and remove any tough strings. Steam or simmer the stalks for 8–10 minutes, or until tender, then drain well and toss in melted butter or olive oil. The stalks can also be blanched, then baked into a gratin with cream, nutmeg and cheese, and they can also be used in risottos and pasta sauces along with the leaves.

Sautéed silverbeet

SERVES 4–6 AS A SIDE DISH

1 kg (2 lb 4 oz) silverbeet (Swiss chard)
2 tablespoons olive oil
3 garlic cloves, finely sliced
extra virgin olive oil, to serve

Trim the leaves from the silverbeet stalks and rinse them in cold water. Blanch the leaves in a large saucepan of boiling salted water for 1–2 minutes, or until tender but still firm. Drain well in a colander, lay out on a tea towel, then, using your hands, gently wring out any excess water.

Heat the oil in a heavy-based frying pan and cook the garlic over low heat until just starting to turn golden. Add the silverbeet, season with sea salt and freshly ground black pepper, then increase the heat to medium and cook for 3–4 minutes, or until warmed through. Transfer to a serving plate and drizzle with extra virgin olive oil, to serve.

Warm silverbeet and chickpea salad

SERVES 4

250 g (9 oz/1 cup) dried chickpeas
1.5 kg (3 lb 5 oz) silverbeet (Swiss chard)
2 tomatoes
125 ml (4 fl oz/$\frac{1}{2}$ cup) olive oil
1 onion, cut into thin wedges
1 teaspoon sugar
$\frac{1}{4}$ teaspoon ground cinnamon
2 garlic cloves, chopped
3 tablespoons chopped mint
2–3 tablespoons lemon juice
1$\frac{1}{2}$ tablespoons sumac

Put the dried chickpeas in a large bowl, cover with water and leave to soak overnight. Drain and place in a large saucepan. Cover with fresh water and bring to the boil, then simmer for 1$\frac{3}{4}$ hours, or until the chickpeas are tender. Drain and set aside.

Meanwhile, thoroughly wash the silverbeet and pat dry with paper towels. Trim the stalks and finely shred the leaves. Cut the tomatoes in half, scrape out the seeds with a teaspoon and dice the flesh. Set aside.

Heat the olive oil in a large heavy-based frying pan, add the onion and cook over low heat for 5 minutes, or until the onion is softened and just starting to brown. Add the tomato to the pan along with the sugar, cinnamon and garlic, and cook for a further 2–3 minutes.

Add the chickpeas and silverbeet to the tomato mixture and cook for 3–4 minutes, or until the silverbeet wilts. Add the mint, lemon juice and sumac, season, and cook for 1 minute. Serve immediately.

Silverbeet is a member of the *Beta vulgaris* species, and is related to beetroot and spinach. One theory as to how it got its alternative name of Swiss chard suggests that it was named after a Swiss botanist who first gave the plant its scientific name. Another theory suggests that, after a plague of flea beetles attacked the chard crop, it was named for its resulting resemblance to the holes in Swiss cheese.

Spanakorizo

SERVES 6 AS A SIDE DISH OR LIGHT MEAL

400 g (14 oz) silverbeet (Swiss chard)
   (see note)
2 tablespoons olive oil
1 large onion, chopped
2 garlic cloves, crushed
6 spring onions (scallions), finely chopped
330 g (11 1/2 oz/1 1/2 cups) white rice
   (short- or medium-grain rice)
1 tablespoon chopped dill
1 tablespoon chopped flat-leaf (Italian)
   parsley
2 tablespoons lemon juice
375 ml (13 fl oz/1 1/2 cups) vegetable stock

Wash the silverbeet well, then cut off the thick stalks and roughly chop the leaves. Blanch the leaves in a large saucepan of boiling salted water. Remove from the pan, rinse under cold water, then drain well.

Heat the olive oil in a large heavy-based saucepan with a tight-fitting lid. Add the onion and sauté over medium heat for 5 minutes, or until softened. Stir in the garlic and cook for a further minute.

Add the spring onion and rice and cook for 2 minutes, stirring constantly to coat the rice. Add the silverbeet, dill, parsley and half the lemon juice. Season well with sea salt and freshly ground black pepper. Stir in the stock and 375 ml (13 fl oz/1 1/2 cups) water. Cover, bring to the boil, then reduce the heat and simmer for 15 minutes.

Remove the pan from the heat and set aside, covered, for 5 minutes. Stir in the remaining lemon juice, then adjust the seasoning if needed, and serve.

note English spinach can be used instead of the silverbeet. Rinse under cold water and add to the rice with the herbs.

spinach

Spinach is thought to have originated in ancient Persia, and in the Arab world today, spinach features in myriad permutations — with eggs and herbs in thick omelettes, added to lamb stews, eaten in conjunction with pulses or enjoyed as a side dish with pine nuts and raisins.

Buying and storing

- Although spinach is available year-round, it is best from mid-autumn to spring.

- Choose bunches of spinach with broad, spade-shaped leaves and undamaged stalks. Baby English spinach leaves can also be bought either loose or in pre-packed bags. Choose small, dry leaves with no bruising or wilting as they are extremely perishable.

- Spinach reduces quite a lot in volume when cooked, so be sure to buy plenty. As a guide, two bunches (about 1 kg/2 lb 4 oz) of spinach will serve four to six people.

- Spinach is easily bruised and won't store for long. Keep for up to 2 days in a plastic bag in the fridge.

Preparation

Spinach needs to be washed carefully in several changes of cold water to remove the dirt, then remove the leaves and discard the thick stalks. The spinach can then be cooked in a covered pan, with only the water from washing still clinging to the leaves, for 3–4 minutes until just wilted. Transfer to a colander and use a large spoon to press down on the spinach to rid it of as much water as possible. Serve as is, or return to the pan with a little butter or cream, and perhaps some freshly grated nutmeg.

To make spinach purée, simply process the cooked spinach in a food processor, season well, then serve as is, or cook it in a little cream or with some butter.

For a simple Japanese-style side dish, dress cooled, wilted baby spinach leaves with soy sauce, rice vinegar and sesame oil, then sprinkle with toasted sesame seeds. Serve with yakitori chicken or barbecued salmon skewers.

Spinach and feta triangles

MAKES 8

1 kg (2 lb 4 oz) English spinach
3 tablespoons olive oil
1 onion, chopped
10 spring onions (scallions), sliced
1 handful parsley, chopped
1 tablespoon chopped dill
a pinch of ground nutmeg
35 g (1¼ oz/⅓ cup) grated parmesan
   cheese
150 g (5½ oz/1 cup) crumbled feta cheese
90 g (3¼ oz/⅓ cup) ricotta cheese
4 eggs, lightly beaten
40 g (1½ oz) butter, melted
1 tablespoon olive oil, extra
12 sheets ready-made filo pastry

Trim any thick stalks from the spinach. Wash the leaves thoroughly, roughly chop and place in a large frying pan with just a little water clinging to the leaves. Cover and cook gently over low heat for 5 minutes, or until the leaves have wilted. Drain well and allow to cool slightly before squeezing the leaves tightly to remove the excess water.

Heat the olive oil in a heavy-based frying pan. Add the onion and cook over low heat for 10 minutes, or until golden. Add the spring onion and cook for a further 3 minutes, then remove from the heat. Stir in the drained spinach, parsley, dill, nutmeg, parmesan, feta, ricotta and egg. Season with sea salt and freshly ground black pepper.

Preheat the oven to 180°C (350°F/Gas 4). Grease two baking trays. Combine the melted butter with the extra olive oil. Work with 3 sheets of pastry at a time, keeping the rest covered with a damp tea towel. Brush each sheet of filo with the butter mixture and lay them on top of each other, then cut the pile of sheets in half lengthways.

Working with one pastry strip at a time, and starting at the end closest to you, spoon 4 tablespoons of the spinach filling on an angle in the corner of the pastry. Fold this corner over the filling to form a triangle. Continue folding the triangle over until you reach the end of the pastry strip. Repeat with all the pastry strips and filling. Put the triangles on the baking trays and brush with the remaining butter mixture. Bake for 20–25 minutes, or until golden brown.

Spinach contains iron and vitamins A and C, but also oxalic acid, which is responsible for its slightly bitter taste, and which acts as an inhibitor to the body's ability to absorb calcium and iron. This knowledge has somewhat diminished its famous 'Popeye' reputation.

Spinach and olive pizza

MAKES TWO 30 CM (12 INCH) PIZZAS

pizza dough
1 tablespoon caster (superfine) sugar
2 teaspoons active dried yeast or
    15 g ($\frac{1}{2}$ oz) fresh yeast
210 ml (7$\frac{1}{2}$ fl oz) lukewarm water
450 g (1 lb/3$\frac{2}{3}$ cups) plain (all-purpose)
    flour
3 tablespoons olive oil
cornflour (cornstarch), to dust

topping
4 tablespoons olive oil
4 garlic cloves, crushed
4 tablespoons pine nuts
2 kg (4 lb 8 oz) English spinach, trimmed
    and roughly chopped
400 ml (14 fl oz) ready-made tomato
    pasta sauce
440 g (15$\frac{1}{2}$ oz/3 cups) grated
    mozzarella cheese
30 very small black olives
50 g (1$\frac{3}{4}$ oz/$\frac{1}{2}$ cup) grated
    parmesan cheese

To make the pizza dough, put the sugar and yeast in a bowl and stir in 90 ml (3 fl oz) of the lukewarm water. Leave in a draught-free place for 10 minutes, or until foamy.

Mix the flour and $\frac{1}{4}$ teaspoon salt in a large bowl and make a well in the centre. Add the olive oil, the remaining lukewarm water and the yeast mixture to the well and mix using a wooden spoon until a rough dough forms. Transfer to a lightly floured surface and knead for 8 minutes, adding a little flour or warm water if necessary, until the dough is soft, smooth and elastic and not sticky to touch.

Place the dough in a large oiled bowl, turning to coat the dough in the oil. Cut a shallow cross on the top with a sharp knife, then cover the bowl with plastic wrap and leave in a draught-free place for 1–1$\frac{1}{2}$ hours, or until doubled in size.

Preheat the oven to 240°C (475°F/Gas 8) and lightly oil two 30 cm (12 inch) round or rectangular baking trays.

To make the topping, heat the olive oil in a large frying pan, add the garlic and pine nuts and fry over low heat, stirring often, for 5–6 minutes, or until golden. Add the spinach in batches if necessary, increase the heat and stir until wilted. Season with salt and freshly ground black pepper.

Gently punch down the dough using a lightly floured fist, then divide in half. Working with one portion at a time, push the dough out to fit the trays.

Dust each pizza base with cornflour and spoon half the tomato sauce onto each base, spreading the sauce just up to the edge. Sprinkle with half the mozzarella. Spread the spinach mixture and olives over the top, then sprinkle with the remaining mozzarella and the parmesan.

Bake for 12–15 minutes, or until the crust is golden and puffed. Brush the rim with a little extra olive oil before serving.

Spinach goes well with anchovies, cheese, yoghurt, cream, eggs, garlic, ginger, nutmeg, paprika, pasta, lemon juice, olive oil, mushrooms, ham, prosciutto, sesame seeds and walnuts.

Roast beef and spinach salad with horseradish cream

SERVES 4

horseradish cream
125 g (4 oz/^{1}/2 cup) Greek-style yoghurt
1 tablespoon bottled horseradish
2 tablespoons lemon juice
2 tablespoons pouring (whipping) cream
2 garlic cloves, crushed
2–3 dashes of Tabasco sauce

200 g (7 oz) green beans, trimmed
500 g (1 lb 2 oz) rump steak, cut into slices
    3 cm (1^1/4 inch) thick
1 red onion, cut in half
1 tablespoon olive oil
100 g (3^1/2 oz/2 cups) baby English
    spinach
50 g (1^3/4 oz) watercress sprigs
200 g (7 oz) semi-dried (sun-blushed)
    tomatoes

Put all the ingredients for the horseradish cream in a small bowl. Add some freshly ground black pepper and whisk together well. Cover and refrigerate for 15 minutes, or until required.

Meanwhile, bring a small saucepan of salted water to the boil. Add the beans and cook for 4 minutes, until tender. Drain, refresh under cold water, then drain well and set aside.

Preheat a barbecue grill or chargrill pan to high. Brush the steak slices and onion halves with the olive oil. Grill the steak for 2 minutes on each side, or until rare. Remove to a plate, cover loosely with foil and set aside to rest for 5 minutes.

Grill the onion for 2–3 minutes on each side, or until lightly charred and tender.

Toss the spinach, watercress, tomatoes and beans together in a large salad bowl. Slice the beef thinly across the grain, then arrange over the salad. Slice the grilled onion, add to the salad and drizzle with the horseradish cream. Season well with sea salt and freshly ground black pepper.

note This cooking time will result in rare beef. Cook it for a little longer if you prefer your beef medium or well done.

Cheese and spinach pancakes

SERVES 4

250 g (9 oz) cooked English spinach,
squeezed dry and chopped (about
600 g/1 lb 5 oz uncooked weight;
see note)
125 g (4$^1/_2$ oz/$^1/_2$ cup) fresh curd cheese
(goat's, sheep's or cow's milk)
30 g (1 oz/$^1/_4$ cup) grated cheddar cheese
a pinch of ground nutmeg
25 g ($^3/_4$ oz/$^1/_4$ cup) grated parmesan cheese
$^1/_2$ teaspoon paprika
40 g (1$^1/_2$ oz/$^1/_2$ cup) fresh breadcrumbs

pancake batter
125 g (4$^1/_2$ oz/1 cup) plain (all-purpose)
flour
310 ml (10$^3/_4$ fl oz/1$^1/_4$ cups) milk
1 egg, lightly beaten
butter, for cooking

cheese sauce
40 g (1$^1/_2$ oz) butter
30 g (1 oz/$^1/_4$ cup) plain (all-purpose) flour
435 ml (15 fl oz/1$^3/_4$ cups) milk
125 g (4$^1/_2$ oz/1 cup) grated cheddar cheese

Combine the spinach, curd cheese, cheddar
and nutmeg in a bowl. Season with freshly
ground black pepper and mix well.

To make the pancake batter, sift the flour and
a pinch of salt into a bowl. Add half the milk
and the egg. Whisk until smooth, then whisk
in the remaining milk.

Heat a teaspoon of butter in a frying pan over
medium heat and slowly pour in enough
batter to create a thin layer in the base of the
pan. Cook for 2–3 minutes, or until golden,
then turn over and cook the other side.
Repeat to make 8 pancakes in total.

Preheat the oven to 180°C (350°F/Gas 4).
Grease a large ovenproof dish.

To make the cheese sauce, melt the butter
over low heat, add the flour and cook for
1 minute. Stirring constantly over medium
heat, gradually add the milk, bringing the
mixture back to a simmer between additions
and stirring well to prevent lumps from
forming. Cook until the mixture thickens, then
remove from the heat, season to taste with
salt and freshly ground black pepper and stir
in the grated cheese.

Divide the spinach filling among the pancakes,
roll the pancakes up and place in the dish,
seam side down. Pour the cheese sauce
over the pancakes. Combine the parmesan,
paprika and breadcrumbs in a bowl, then
sprinkle over the sauce. Bake for 30 minutes,
or until golden brown. Serve immediately.

note You can use 250 g (9 oz) frozen
spinach if preferred. Allow it to thaw, then
squeeze out all the water.

witlof

In the same family as radicchio and frisée, witlof (also called chicory or Belgian endive) is a slightly bitter vegetable with compact, slender heads made up of yellow- or maroon-tipped cream leaves. Expensive and labour-intensive to harvest, witlof is grown in the dark to prevent sunlight from turning the leaves from white to green.

Buying and storing

- The peak season for witlof is from mid-autumn to early spring.

- Witlof is sensitive to light, which makes it bitter when exposed to it; you'll find it is usually sold wrapped in paper to prevent this. Look for crisp, tightly furled heads with pale yellow or reddish maroon tips, as green chicory will be quite bitter.

- The longer witlof is stored, the more bitter it will become. Wrap in paper towel or the paper it was sold in and place inside a plastic bag, then store for no more than 1 day in the fridge.

Preparation

Cut off the base, then separate the leaves and wash under cold running water. Do not soak witlof, as this can make it turn bitter, and don't cook in cast-iron pans, as this will turn the leaves a shade of grey.

Braising Cut heads of witlof in half and pack snugly into an ovenproof dish. Add enough hot chicken stock to cover to halfway, cover with foil and cook in a 180°C (350°F/Gas 4) oven for 40 minutes, or until tender. Sprinkle with blue cheese, drizzle with cream and cook for a few more minutes until heated through. The braised tender witlof can also be barbecued or chargrilled briefly until browned.

Witlof goes well with ingredients such as blue cheese, butter, cream, red meat and game meat, olives and orange.

Sautéing Cut the witlof into large chunks and toss in a large pan with plenty of butter over medium–high heat for 5–6 minutes, or until softened and golden.

Witlof and radicchio bake

SERVES 4 AS A SIDE DISH

450 g (1 lb) white witlof (chicory/Belgian endive)
1 large radicchio
150 g (5 1/2 oz) pancetta or smoked bacon, thinly sliced
55 g (2 oz/2/3 cup) fresh breadcrumbs
50 g (1 3/4 oz/1/2 cup) grated parmesan cheese
1 1/2 tablespoons finely chopped thyme
1 garlic clove, finely chopped
560 ml (19 fl oz/2 1/4 cups) thick (double/heavy) cream

Preheat the oven to 180°C (350°F/Gas 4). Slice the witlof in half lengthways (or if they are quite large, slice them into quarters). Divide the radicchio into six or eight wedges, depending on its size.

Lightly butter a shallow 2.5 litre (87 fl oz/10 cup) gratin dish. Place the witlof and radicchio in the dish in one layer, alternating the colours.

Mix together the pancetta, breadcrumbs, parmesan, thyme and garlic and season well. Sprinkle over the witlof and radicchio.

Pour the cream over the top, cover with foil and bake for 50–60 minutes. Take the foil off the dish for the last 20 minutes to crisp up the pancetta and breadcrumbs. Leave to rest for about 10 minutes before serving.

Sautéed witlof with olives and caperberries

SERVES 4 AS A STARTER OR SIDE DISH

40 g (1 1/2 oz/1/4 cup) pitted kalamata olives, chopped
2 anchovy fillets, drained and chopped
6 small caperberries
1 tablespoon olive oil
2 white or red witlof (chicory/Belgian endive)
20 g (3/4 oz) unsalted butter
1 garlic clove, crushed
a pinch of chilli flakes (optional)

Put the olives and anchovies in a small bowl. Chop two of the caperberries and add to the olives along with half the olive oil. Stir to combine well.

Discard the outer leaves from each witlof and cut the heads in half lengthways. Open out the leaves a little and spoon the olive mixture between the leaves. Join the two halves together again and tie with kitchen string to secure.

Heat the remaining oil and the butter in a saucepan over low heat. Add the witlof, garlic and chilli flakes, if using, then cover and cook for 8–10 minutes, turning the witlof once and adding a little hot water if necessary to stop it sticking.

To serve, untie the string and divide the witlof, cut side up, among serving plates. Spoon over any pan juices. Slice the remaining caperberries in half lengthways and scatter them over the witlof. Serve hot.

zucchini

Despite their Italian name, these members of the summer squash family are Mexican natives. The word zucchini is from the Italian *zuccherino* meaning 'small squash'; the French name courgette, also used in England, is from *courge*, for marrow, and 'squash' is from the American Indian word *askutasquash*, which means 'green thing eaten raw'.

Buying and storing

- Although they are sold year-round, zucchini are at their peak during summer.

- Zucchini come in green, yellow and a so-called white variety, but this is really pale green and is more bulbous than the others.

- Choose zucchini with firm, shiny skins. The skin is very delicate, so a few scratches from harvesting and transporting are inevitable. The flesh should be creamy and bright and have the tiniest of seeds.

- Buy and eat zucchini quickly as they don't keep for longer than 2 days; storage will affect their delicate flavour and texture. Store in a ventilated bag in the crisper of the fridge, and don't wash them until you are ready to cook them or they may start to spoil.

Preparation

Zucchini simply require washing and their stem-ends trimmed, then they can be sliced, grated, chopped or halved lengthways. They cook quickly and are suited to steaming, stir-frying, baking, sautéing in olive oil (add some garlic and fresh thyme), braising or barbecuing. If frying them or using in fritters, salt them first to degorge them so they soak up less oil.

To chargrill zucchini, slice them lengthways into strips, brush with olive oil and cook for about 1 minute on each side on a hot barbecue grill or chargrill pan. You can then combine them with other chargrilled vegetables such as tomatoes, eggplant (aubergine) and red and green capsicums (peppers).

Zucchini can also be served raw, as the Italians do, thinly sliced and dressed with olive oil, lemon juice and mint, and seasoned with sea salt and freshly ground black pepper.

Chargrilled vegetable terrine

SERVES 8

8 large slices chargrilled eggplant (aubergine)
(see note)
10 slices chargrilled red capsicum (pepper)
8 slices chargrilled zucchini (courgette)
350 g (12 oz/1$\frac{1}{3}$ cups) ricotta cheese
2 garlic cloves, crushed
2 handfuls rocket (arugula)
3 marinated artichokes, drained and
sliced
85 g (3 oz/$\frac{1}{2}$ cup) semi-dried (sun-blushed)
tomatoes, drained and chopped
100 g (3$\frac{1}{2}$ oz) marinated mushrooms,
drained and halved

Line a 23 x 13 x 6 cm (9 x 5 x 2$\frac{1}{2}$ inch) loaf
(bar) tin with plastic wrap, leaving a generous
overhang on each side.

Line the base of the tin with half of the
eggplant, cutting it and patching it to fit
evenly. Next, lay half the capsicum evenly
over the top of the eggplant, then layer all
of the zucchini.

Using a wooden spoon, beat the ricotta and
garlic together in a bowl until smooth. Season
with sea salt and freshly ground black pepper,
then spread the cheese evenly over the
zucchini, pressing down firmly. Top with the
rocket leaves. Arrange the artichoke, tomato
and mushrooms over the rocket. Top with
another layer of capsicum and finish with the
remaining eggplant.

Cover the terrine, using the overhanging
plastic wrap. Place a piece of cardboard on
top of the terrine, then use tins of food to
press it down. Refrigerate overnight.

Peel back the plastic wrap and carefully turn
the terrine out onto a platter. Remove the
plastic wrap, cut into thick slices and serve.

note You can chargrill the vegetables
yourself or use ready-made chargrilled
vegetables sold from delicatessens. These are
usually sitting in oil and need to be drained
before use.

Zucchini represent just one of the many 'summer' squashes
available — these have softer skin and watery flesh, while winter
squash, such as pumpkins, have hard skin and flesh. **Pattypan
squash**, with its flat, round shape and scalloped edge, are picked
when young and are prized for their buttery soft flesh. These
summer squash are usually a distinctive bright yellow colour, but
some are either dark or light green. Choose small firm squash
and steam them whole, for 8–10 minutes, or until tender.

Zucchini patties

SERVES 4 AS A STARTER OR SIDE DISH

cucumber and yoghurt salad
1 Lebanese (short) cucumber
250 g (9 oz/1 cup) Greek-style yoghurt
1 small garlic clove, crushed
1 tablespoon chopped dill
2 teaspoons white wine vinegar
white pepper, to season

300 g (10 oz) zucchini (courgettes), grated
1 small onion, finely chopped
3 tablespoons self-raising flour
4 tablespoons grated kefalotyri or parmesan cheese
1 tablespoon chopped mint
2 teaspoons chopped parsley
a pinch of ground nutmeg
3 tablespoons dry breadcrumbs
1 egg, lightly beaten
olive oil, for pan-frying
rocket (arugula) leaves, to serve
lemon wedges, to serve (optional)

To make the cucumber and yoghurt salad, chop the cucumber into small pieces, then place in a colander, sprinkle with salt and set aside in the sink or on a plate to drain for 15–20 minutes.

In a bowl, mix together the yoghurt, garlic, dill and vinegar. Add the cucumber and season to taste with salt and white pepper. Cover and refrigerate until required.

Meanwhile, preheat the oven to 120°C (235°F/Gas 1/2).

Put the zucchini and onion in a clean tea towel, gather the corners together and twist as tightly as possible to remove all the juices. Tip the zucchini and onion into a large bowl, then add the flour, cheese, mint, parsley, nutmeg, breadcrumbs and egg. Season well with salt and freshly ground black pepper, then mix with your hands to a stiff batter.

Heat 1 cm (1/2 inch) olive oil in a large heavy-based frying pan over medium heat. When the oil is hot, drop 2 tablespoons of the batter into the pan and press flat to make a thick patty. Fry several patties at a time for 2–3 minutes, or until well browned all over. Drain well on paper towels and place in the oven to keep warm while cooking the remaining patties.

Serve the zucchini patties hot with rocket leaves and the cucumber and yoghurt salad. Serve with lemon wedges for squeezing over, if desired.

Serve **zucchini** with the robust flavours of garlic, ginger, balsamic and red wine vinegars, lemon and anchovies, as well as classic Mediterranean ingredients like oregano, mint, basil, tomatoes, pine nuts and parmesan cheese.

Zucchini with lemon and caper butter

SERVES 4 AS A SIDE DISH

lemon and caper butter
100 g (3 1/2 oz) butter, softened
2 tablespoons capers, rinsed, squeezed dry
    and chopped
2 teaspoons finely grated lemon zest
1 tablespoon lemon juice

750 g (1 lb 10 oz) small zucchini
    (courgettes) (about 8), trimmed

To make the lemon and caper butter, put the butter, capers, lemon zest and juice in a large bowl. Using a wooden spoon, beat until well combined, then season with salt and freshly ground black pepper.

Thinly slice the zucchini lengthways. Steam over boiling water for 3–4 minutes, or until just tender, then toss with the lemon and caper butter in a bowl and serve immediately.

Deep-fried zucchini flowers

SERVES 4 AS A SIDE DISH

2 eggs
60 g (2 1/4 oz/1/2 cup) plain (all-purpose)
    flour
10–12 zucchini (courgette) flowers (see note)
oil, for deep-frying

stuffing
125 g (4 1/2 oz) ricotta cheese
1 tablespoon chopped basil
2 tablespoons grated parmesan cheese
2 tablespoons fresh breadcrumbs
1 egg yolk

Whisk the eggs with the flour and season with salt and freshly ground black pepper.

To make the stuffing for the flowers, combine the ricotta, basil, parmesan, breadcrumbs and egg yolk in a bowl and stir to mix well. Season with salt and pepper, then use the mixture to stuff the flower cavities.

Half-fill a medium saucepan with oil. Heat the oil to 180°C (350°F), or until a cube of bread dropped into the oil turns golden brown in 15 seconds.

Add a little cold water to the batter if it is too thick, then, working in batches, dip each stuffed zucchini flower into the batter, draining excess batter, and deep-fry for 4-5 minutes, or until golden. Drain the flowers well on paper towels and serve immediately.

note Before using zucchini flowers, remove and discard the stamen from inside each flower, wash the flower and make sure it doesn't harbour any insects. The flowers only last a few days after being picked, so it is best to buy them on the day you plan to use them.

Zucchini flowers are edible too, available in the male (the flower has a stalk) and the female (the flower has a baby zucchini attached to it) form. Brush them gently to remove their furry down covering and gently rinse to remove any insects hidden inside. The flowers can be lightly battered and fried, stuffed and baked, cooked into a frittata, or sliced and stirred through pastas and risottos. Use on the day of purchase or store, covered, in a damp cloth-lined dish in the fridge for a day or so.

Pappardelle with zucchini flowers and goat's cheese

SERVES 4

175 g (6 oz/^{3}/4 cup) ricotta cheese
125 ml (4 fl oz/^{1}/2 cup) thick (double/heavy) cream
2 teaspoons thyme
^{1}/4 teaspoon ground nutmeg
white pepper, to season
300 g (10^1/2 oz) dried pappardelle pasta or 400 g (14 oz) fresh pappardelle or other ribbon pasta
3 tablespoons olive oil
4 small zucchini (courgettes), cut into thin batons
16 zucchini (courgette) flowers, no vegetable attached
plain (all-purpose) flour, to dust
100 g (3^1/2 oz/^{2}/3 cup) crumbled goat's cheese

Combine the ricotta, cream, thyme and nutmeg in a bowl and season well with sea salt and white pepper. Set aside in a cool place (do not refrigerate) for 1 hour.

Cook the pasta in a large saucepan of boiling salted water until al dente. Meanwhile, heat the olive oil in a large frying pan over medium heat and cook the zucchini for 4 minutes, or until lightly golden. Remove the zucchini from the pan with a slotted spoon and drain on paper towels.

Dust the zucchini flowers with flour, shake off the excess and fry for about 1 minute, or until lightly golden.

Drain the pasta and transfer to a large serving dish. Add the zucchini, zucchini flowers and ricotta mixture and toss lightly. Dot with small pieces of goat's cheese, to serve.

salad vegetables

avocado

Indigenous to Mexico and tropical Central America, the avocado is an incredible vegetable — although botanically it is classed as a fruit. Avocados have soft buttery flesh and a mild, slightly nutty flavour, and they are the only fruit to contain fat. This fat is mainly monounsaturated ('good') fat, and can actually help lower certain types of cholesterol in the body.

Varieties

Hass This popular avocado has knobbly green skin, which turns inky-purple when fully ripe and has rich, green buttery flesh. The skin can be difficult to remove. This is the only variety that grows year-round.

Sharwill These have knobbly green skin and rich, creamy flesh. This variety is available from late autumn to spring.

Fuerte These smooth-skinned avocados have a creamy, pale yellow flesh. Make sure they are fully ripe or their subtle flavour can be a little bland. The best time for these is in autumn.

Reed These large rounded fruit have thin, shiny easy-to-peel skin, and pale yellow buttery flesh. They are available during summer.

Buying and storing

- To check if the avocado is fully ripe, hold the fruit in your hands and apply slight pressure — it should 'give' slightly.

- Unripe avocados are hard and tasteless, but will take only a few days to soften if stored at room temperature. When ripe, store in the fridge and use within 1–2 days.

Preparation

Avocados are best eaten raw as heat brings out their bitter tannins. To prepare the avocado, cut it in half lengthways around the large central stone. Rotate the halves to separate them, then remove the stone with the tip of a spoon. Cut avocado turns brown, so cut it just before use or brush with lemon juice to stop it discolouring.

The skin of some avocados can be brittle or difficult to peel off. If so, it may be easier to use a large spoon to scoop the flesh out of its skin, or to slice or dice the flesh while it is still in its skin, then scoop the pieces out.

Avocado salsa

SERVES 6

1 red onion
2 large avocados
1 tablespoon lime juice
1 tomato
1 small red capsicum (pepper)
1 teaspoon ground coriander
1 teaspoon ground cumin
3 tablespoons chopped coriander (cilantro)
2 tablespoons olive oil
4–5 drops Tabasco sauce
corn chips, to serve

Finely chop the onion. Cut the avocados in half, remove the stones and cut the flesh into cubes while still in the skin. Use a large spoon to scoop the cubes into a bowl and toss lightly with lime juice.

Cut the tomato in half widthways, squeeze gently to remove the seeds, then finely chop the flesh. Remove the seeds and membrane from the capsicum and finely dice. Add to the bowl with the avocado.

Put the ground coriander and cumin in a small frying pan over medium heat. Stir the spices for 1 minute to enhance their flavour and fragrance. Allow to cool, then add to the bowl along with the olive oil and Tabasco sauce. Stir gently to combine, being careful not to squash the avocado. Refrigerate until required and serve at room temperature with corn chips.

Avocado, bacon and tomato salad

SERVES 2

3 garlic cloves, unpeeled
2 tablespoons olive oil
3 teaspoons balsamic vinegar
1 teaspoon dijon mustard
125 g (4 1/2 oz) bacon slices
1 avocado
50 g (1 3/4 oz) mixed green salad leaves
1/2 small red onion, thinly sliced
2 small tomatoes, cut into chunks

Preheat the oven to 180°C (350°F/Gas 4). Put the unpeeled garlic cloves on a baking tray and roast for 30 minutes. Remove, allow to cool, then squeeze the garlic flesh out of the skins and mash in a small bowl. Add the olive oil, vinegar and mustard, then whisk to combine well.

Chop the bacon into bite-sized pieces, then cook under a medium–hot grill (broiler) or dry-fry in a frying pan over medium heat for 3–5 minutes, or until crisp.

Cut the avocado in half, remove the stone and cut the flesh into large cubes while still in the skin. Use a large spoon to scoop the cubes into a serving bowl. Add the bacon, salad leaves, onion and tomato and gently toss to combine. Drizzle the garlic dressing over the salad just before serving.

capsicum

Also known as pepper, bell pepper and sweet pepper, capsicums are botanically a fruit but they are treated more as a vegetable or salad ingredient. Fleshy, crisp and sweet (except for the green ones), capsicums feature in many of the world's great ethnic cuisines, including Italian, Spanish, Chinese and Mexican.

Buying and storing

- The peak season for capsicums is from summer to mid-autumn, although they are available year-round.

- Most capsicums are green at first and then turn red, yellow or orange or even purple-black, depending on the variety. Those with thick flesh are the juiciest, and red capsicums are generally sweeter than green ones.

- Buy capsicums that are heavy for their size and that have bright, tight, smooth, glossy skin with no wrinkled patches.

- Store capsicums in a ventilated bag in the crisper of the fridge for up to 1 week.

Preparation

To prepare capsicums, slice off the stem, cut them in half, trim away the white membrane (this will be bitter) and scrape out the seeds.

Peeling Capsicums can be used peeled or unpeeled, but if you find the skin difficult to digest, remove the skin from a whole capsicum with a potato peeler. Slice off the fleshy sides, leaving just the seed-filled core and hard stem end. Slice, dice or chop the flesh and use as desired. The skin can also be removed by chargrilling or roasting the capsicum first.

Chargrilling Roasting or grilling capsicums makes them sweeter and also gives them a smoky flavour if you use a chargrill or barbecue. Place the whole capsicums on a chargrill pan or barbecue grill and cook for 10 minutes or so, turning frequently, until the skin blackens and blisters. Cool them in a paper bag or in a large bowl covered with a tea towel, then pull off all the black skin. Don't wash off any skin or seeds as the water will only dilute the flavour. You can also do this under a grill (broiler), using either whole or halved capsicums.

Roasting Brush whole capsicums lightly with oil, place in a roasting tin and cook in a 220°C (425°F/Gas 7) oven for 15–20 minutes, turning occasionally, or until the flesh has softened and the skin looks burnt. Cool (as for chargrilled ones), then peel off the skin, remove the stem, membrane and seeds, and use as required.

Stuffed vegetables provençale

SERVES 6 AS A STARTER OR SIDE DISH

2 small eggplants (aubergines), halved
    lengthways
2 small zucchini (courgettes), halved
    lengthways
4 tomatoes
2 small red capsicums (peppers)
4 tablespoons olive oil
2 red onions, finely chopped
2 garlic cloves, crushed
250 g (9 oz) minced (ground) pork
250 g (9 oz) minced (ground) veal
3 tablespoons tomato paste (concentrated
    purée)
4 tablespoons white wine
2 tablespoons chopped parsley
50 g (1 3/4 oz/1/2 cup) grated parmesan
    cheese
80 g (2 3/4 oz/1 cup) fresh breadcrumbs
extra virgin olive oil, to serve
crusty bread, to serve

Preheat the oven to 180°C (350°F/Gas 4).
Grease a large roasting tin with olive oil.

Use a spoon to hollow out the centres of the
halved eggplants and zucchini, leaving a
border around the edge. Chop the eggplant
and zucchini flesh finely and set aside.

Cut the tops off the tomatoes and reserve.
Use a spoon to hollow out the centres,
catching the juice in a bowl, then chop the
flesh roughly. Set the flesh and juice aside.

Cut the tops off the capsicums and reserve.
Discard the seeds and membranes from
inside the capsicum shells. Set the capsicum
shells aside.

Heat half the olive oil in a large frying pan.
Add the onion and garlic and sauté over
medium–high heat for 3 minutes, or until
softened. Add the pork and veal and stir for
5 minutes, or until the meat browns, breaking
up any lumps with the back of a fork.

Stir in the chopped eggplant and zucchini
flesh and cook for 3 minutes, then stir in the
chopped tomato and reserved juice, along
with the tomato paste and wine. Cook, stirring
occasionally, for 10 minutes. Remove from
the heat and stir in the parsley, parmesan
and breadcrumbs. Season well with sea salt
and freshly ground black pepper, then spoon
the mixture into the eggplant, zucchini,
tomato and capsicum shells. Put the tops
back on the tomatoes and capsicums.

Arrange the capsicums and eggplant in a
single layer in the roasting tin. Drizzle with
some of the remaining olive oil, pour 125 ml
(4 fl oz/1/2 cup) water into the roasting tin
and bake for 15 minutes. Then add the
tomatoes to the roasting tin in a single layer
and bake for 5 minutes. Finally add the
zucchini, drizzle with the remaining olive oil
and bake for a further 25 minutes, or until
all the vegetables are tender.

Serve the stuffed vegetables hot or at room
temperature, drizzled with some extra virgin
olive oil, with slices of crusty bread.

Capsicum and bean stew

SERVES 4–6

200 g (7 oz/1 cup) dried haricot beans
  (see note)
2 tablespoons olive oil
1 red onion, halved and cut into thin wedges
2 large garlic cloves, crushed
1 red capsicum (pepper), diced
1 green capsicum (pepper), diced
2 x 400 g (14 oz) tins chopped tomatoes
2 tablespoons tomato paste (concentrated
  purée)
500 ml (17 fl oz/2 cups) vegetable stock
2 tablespoons chopped basil
125 g (4 1/2 oz/3/4 cup) kalamata olives, pitted
1–2 teaspoons soft brown sugar
crusty bread, to serve

Put the beans in a large bowl, cover with cold water and leave to soak overnight. Rinse well, then place in a saucepan, cover with plenty of fresh cold water and bring to the boil. Reduce the heat and simmer for 45 minutes, or until just tender. Drain and set aside.

Heat the olive oil in a large heavy-based saucepan. Add the onion and garlic and sauté over medium heat for 3 minutes, or until the onion has softened a little. Add the capsicums and cook for a further 5 minutes.

Stir in the tomatoes, tomato paste, stock and beans. Cover and simmer for 40 minutes, or until the beans are cooked through. Stir in the basil, olives and sugar, then season to taste. Serve hot, with crusty bread.

note Instead of dried beans you could use two 400 g (14 oz) tins of drained and rinsed haricot beans. Add these at the end of cooking with the basil and olives.

Capsicum rolled with goat's cheese and basil

SERVES 4–6 AS A STARTER OR SIDE DISH

4 large red capsicums (peppers)
1 large handful flat-leaf (Italian) parsley,
  chopped
2 tablespoons snipped chives
2 tablespoons small capers, rinsed, drained
  and finely chopped
1 tablespoon balsamic vinegar
150 g (5 1/2 oz/1 1/4 cups) crumbled soft
  goat's cheese
16 basil leaves
olive oil, to cover
crusty Italian bread, to serve

Preheat the grill (broiler) to high. Place the capsicums on a tray under the grill and cook, turning often, for 10–15 minutes, or until the skin is blackened all over. Put the capsicums in a plastic bag and leave to cool, then peel away the skin. Cut the capsicums into 3 cm (1 1/4 inch) wide strips (you will need about 16 strips).

Combine the parsley, chives, capers and vinegar in a small bowl. Add the goat's cheese and mix well. Season with freshly ground black pepper.

Lay the capsicum pieces, skin side down, on a work surface, then place a basil leaf on each capsicum piece. Top each with a teaspoon of the goat's cheese mixture. Roll each piece of capsicum up and over the goat's cheese, to form a roll. Secure with a toothpick. Place the rolls in an airtight, non-reactive container and cover with olive oil. Refrigerate until required, then bring to room temperature before serving. Serve with crusty Italian bread.

Chilli con queso

SERVES 6–8

30 g (1 oz) butter
1/2 red onion
2 large green chillies, seeded, finely chopped
2 small red chillies, seeded, finely chopped
1 garlic clove, crushed
1/2 teaspoon sweet paprika
1 1/2 tablespoons beer
125 g (4 1/2 oz/1/2 cup) sour cream
200 g (7 oz) cheddar cheese, grated
1 tablespoon chopped coriander (cilantro)
1 tablespoon sliced jalapeños
corn or tortilla chips, to serve

Melt the butter in a saucepan over medium heat. Add the onion and chillies and cook for 5 minutes, or until softened. Increase the heat to high, add the garlic and paprika and cook for 1 minute, or until fragrant.

Add the beer, bring to the boil and cook until almost evaporated. Reduce the heat to low and add the sour cream, stirring until smooth. Add the cheese and stir until the cheese is just melted and the mixture is smooth. Remove from the heat and stir in the coriander and jalapeño, and season to taste. Serve with corn or tortilla chips for dipping.

Pasta with fresh chilli and herbs

SERVES 4

500 g (1 lb 2 oz) long pasta, such as fettucine
125 ml (4 fl oz/1/2 cup) olive oil
5 garlic cloves, finely chopped
3–4 small red chillies, seeded and thinly sliced
4 anchovies, finely chopped
1 large handful flat-leaf (Italian) parsley, roughly chopped
1 small handful oregano, finely chopped
1 small handful basil, chopped
2 tablespoons lemon juice
shaved parmesan cheese, to serve

Bring a large saucepan of salted water to the boil. Cook the pasta until al dente, then drain. Meanwhile, put the olive oil, garlic, chilli and anchovies in a small saucepan over low heat and cook, stirring, for 10 minutes, or until the garlic is lightly golden. Remove from the heat.

Add the oil mixture to the drained pasta along with the parsley, oregano, basil and lemon juice. Toss to combine. Season to taste and serve with shaved parmesan.

Chillies and capsicums are closely related, the main difference being that capsicums contain a recessive gene, which means they don't contain any capsaicin, the compound that makes chillies so hot. But not all chillies provide that distinctive blast of 'heat' — red chillies are hotter than green or yellow ones, and small chillies tend to be hotter than larger ones.

Chilli jam

MAKES 1 LITRE (35 FL OZ/4 CUPS)

500 g (1 lb 2 oz) red capsicums (peppers)
125 g (4 1/2 oz) long red chillies, seeded
310 ml (10 3/4 fl oz/1 1/4 cups) white vinegar
880 g (1 lb 15 oz/4 cups) sugar
185 g (6 1/2 oz/1 cup) soft brown sugar

Preheat the grill (broiler) to high. Cut the capsicums into quarters and remove the seeds and membranes. Grill the capsicums, skin side up, until the skin blackens and blisters. Transfer to a bowl, cover with plastic wrap and leave until cool enough to handle. Slip the blackened skin off the capsicums. Put the capsicum and chilli in a food processor with 3 tablespoons of the vinegar and process until finely chopped.

Put the capsicum and chilli mixture in a large saucepan and add the remaining vinegar. Bring to the boil, then reduce the heat and simmer for 8 minutes. Remove the pan from the heat. Add the sugars and stir until the sugar has dissolved, then return the pan to the heat and boil for 5–10 minutes, or until the mixture has thickened slightly.

Spoon immediately into sterilised, warm glass jars and seal. Turn the jars upside down for 2 minutes, then invert and leave to cool. Leave for 1 month before opening to allow the flavours to develop.

Store the chilli jam in a cool, dark place for up to 12 months. After opening, store in the fridge and use within 6 weeks.

Marinated capsicums

SERVES 6 AS A SIDE DISH

3 red capsicums (peppers)
3 thyme sprigs
1 garlic clove, thinly sliced
2 teaspoons roughly chopped flat-leaf
  (Italian) parsley
1 bay leaf
1 spring onion (scallion), sliced
1 teaspoon paprika
3 tablespoons extra virgin olive oil
2 tablespoons red wine vinegar

Preheat the grill (broiler) to high. Cut the capsicums into quarters and remove the seeds and membranes. Grill the capsicums, skin side up, until the skin blackens and blisters. Transfer to a bowl, cover with plastic wrap and leave until cool enough to handle. Slip the blackened skin off the capsicums, then thinly slice the flesh. Place in a bowl along with the thyme, garlic, parsley, bay leaf and spring onion.

In a small bowl, whisk together the paprika, olive oil, vinegar and some sea salt and freshly ground black pepper. Pour over the capsicum mixture and toss to combine well. Cover and refrigerate for at least 3 hours, or preferably overnight, to allow the flavours to develop.

Serve at room temperature as part of an antipasti selection, or as an accompaniment to grilled meats or fish. Store the capsicums in the fridge, covered, for 3 days.

celery

Buying and storing

- The peak season for celery is from summer to winter, although it is available year-round.

- Buy celery with the leaves attached, as these are a good indicator of freshness; you should not be able to bend the stalks. The 'heart' of a bunch of celery (prized for its delicate flavour and texture) should be pale yellow.

- Wrap in plastic and store in the crisper of the fridge for up to 1 week. Celery is made up of 95 per cent water, so don't keep it in the coldest part of the fridge as it may freeze.

Preparation

Some celery may be stringy: peel with a potato peeler to remove the fibrous layer. Discard the dark green leaves and the first few centimetres of the stalks (or use them in stocks). Use the tough outer stalks for soups, stocks and braises.

Celery is one of those vegetables that most cooks have in their refrigerator, but which invariably ends up being chopped into a soup or sauce, or served with raw sticks of carrot as crudités. Yet celery stands up well to cooking and its uses are many. While it's hard to imagine a waldorf salad without it or a bloody mary without its celery stalk swizzle stick, celery can also be made into a puréed soup (add a potato or two for texture), and trimmed celery stalks can be baked into a gratin with béchamel sauce and blue or gruyère cheese.

Tender **celery hearts**, leaves included, are lovely in salads or salsas, or chopped with capers, lemon, onion and parsley and a dash of olive oil, and served with fish. Finely chopped and folded through some egg mayonnaise with a few capers and some mustard, they make a great sauce to serve with fish or poached chicken.

Braised celery

SERVES 4 AS A SIDE DISH

30 g (1 oz) butter
1 bunch of celery, trimmed and cut into
    5 cm (2 inch) lengths
500 ml (17 fl oz/2 cups) chicken or
    vegetable stock
2 teaspoons finely grated lemon zest
3 tablespoons lemon juice
3 tablespoons cream
2 egg yolks
1 tablespoon cornflour (cornstarch)
a pinch of ground nutmeg
1–2 tablespoons chopped parsley

Preheat the oven to 180°C (350°F/Gas 4). Grease a large shallow baking dish.

Melt the butter in a large frying pan. Add the celery, toss to coat in the butter, then cover and cook over medium heat for 2 minutes. Pour in the stock. Add the lemon zest and juice, then cover and simmer for 10 minutes, or until the celery is tender, but still holds its shape. Remove the celery using a slotted spoon and place in the baking dish. Reserve 3 tablespoons of the cooking liquid.

In a bowl, mix together the cream, egg yolks and cornflour. Whisk in the reserved cooking liquid. Pour the mixture back into the frying pan and cook, stirring constantly, until the mixture boils and thickens. Add the nutmeg and season to taste with sea salt and freshly ground black pepper.

Pour the sauce over the celery and bake for 15 minutes, or until the celery is very soft and the sauce is bubbling. Scatter the parsley over the top.

Serve warm with poached chicken breast, chargrilled lamb or slices of corned beef.

Waldorf salad

SERVES 4

butter lettuce leaves, to serve
2 red apples, quartered and cored
1 large green apple, quartered and cored
1½ celery stalks, sliced
3 tablespoons walnut halves
2 tablespoons ready-made egg mayonnaise
1 tablespoon sour cream

Line a serving bowl with lettuce leaves. Cut the apples into 2 cm (¾ inch) chunks and place in a large mixing bowl with the celery and walnuts.

In a small bowl, combine the mayonnaise and sour cream and mix well. Stir the dressing through the apple, celery and walnut mixture, then transfer to the lettuce-lined serving bowl and serve.

cucumber

Belonging to the same family as melons and gourds, cucumbers exist in boundless varieties, coming in round and oval shapes, yellow, pale green and bronze colours, and various sizes, from tiny gherkins to mammoth types up to 50 cm (20 inches) in length. Their juicy, crunchy texture makes them a refreshing ingredient in salads or mix with yoghurt in Indian raitas.

Varieties

Garden Also called the common cucumber, this variety is stumpy and has rounded ends and thick, slightly bitter skin with large seeds. It is usually waxed, so peel and seed it before use.

Lebanese (short) These juicy, mildly sweet cucumbers are about 10 cm (4 inches) in length. Use in salads or any recipe calling for raw, unpeeled cucumber.

Telegraph (long) Also called the English, continental or 'burpless' cucumber, these long, slim cucumbers with small seeds are usually wrapped in plastic to protect against drying out. Use peeled or unpeeled in salads and salsas. Some are waxed, so these need to be peeled.

Gherkin The classic pickling cucumber, these short, warty-skinned specimens have been bred specifically for this purpose. They are pickled whole and don't need to be peeled before use.

Buying and storing

- Cucumbers are available year-round although their peak season is summer.

- Buy firm, unblemished cucumbers that feel heavy, and choose the smaller ones as these will be sweeter. Inspect the ends for softness, as this is a sign of age or poor storage.

- Store in a ventilated plastic bag in the crisper of the fridge for up to 5 days.

Preparation

Peeling First cut it in half horizontally (if your cucumber is long, it will be easier to handle if you cut it in half first). Peel the cucumber with a potato peeler or a small sharp knife.

Seeding Use a small teaspoon to scrape out the seeds from the two halves, then drain the cucumber with the seeded side down. Cut or use as desired.

Tsatsiki

MAKES 500 G (1 LB 2 OZ/2 CUPS)

2 Lebanese (short) cucumbers
400 g (14 oz/1²/₃ cups) Greek-style yoghurt
4 garlic cloves, crushed
3 tablespoons finely chopped mint
1 tablespoon lemon juice
chopped mint, to garnish

Cut the cucumbers in half lengthways, scoop out the seeds with a teaspoon and discard them. Leave the skin on and coarsely grate the cucumber into a small colander. Sprinkle with a little salt and leave to stand over a large bowl for 15 minutes to drain off any bitter juices.

Meanwhile, stir together the yoghurt, garlic, mint and lemon juice in a bowl. Rinse the cucumber under cold water. Then, taking small handfuls, squeeze out any excess moisture. Combine the cucumber with the yoghurt mixture and season to taste. Serve immediately or refrigerate until ready to serve, then garnish with some chopped mint. Serve as a dip with flatbreads or as a sauce for seafood and meat dishes.

Greek salad

SERVES 4

1 large tomato
1 Lebanese (short) cucumber
2 radishes
1 small onion
100 g (3¹/₂ oz) feta cheese
45 g (1¹/₂ oz/¹/₃ cup) pitted black olives
2 tablespoons lemon juice
3 tablespoons olive oil
¹/₂ teaspoon oregano
coral lettuce, to serve (optional)

Chop the tomato into large cubes and cut the cucumber into rounds. Thinly slice the radishes and onion, and cut the feta cheese into small cubes.

Combine the prepared tomato, cucumber, radishes, onion, feta and olives in a serving bowl. Drizzle with the combined lemon juice, olive oil and oregano. Serve on a bed of coral lettuce, if you like.

Cucumber goes well with herbs such as mint, parsley, chives and tarragon, and other ingredients such as chilli, ginger, lemon, butter, feta cheese, goat's cheese, yoghurt, cream and sour cream. Sliced or grated cucumbers release a lot of water upon standing, so discard the water or prepare the cucumbers at the last moment before adding them to a yoghurt- or sour-cream based dish.

Barbecued salmon cutlets with sweet cucumber dressing

SERVES 4

sweet cucumber dressing
2 small Lebanese (short) cucumbers, peeled, seeded and finely diced
1 red onion, finely chopped
1 red chilli, finely chopped
2 tablespoons pickled ginger, shredded (see note)
2 tablespoons rice vinegar
$1/2$ teaspoon sesame oil

oil, for brushing
4 salmon (or ocean trout) cutlets
1 sheet toasted nori (dried seaweed), cut into thin strips (see note)
steamed rice, to serve

In a bowl, mix together all the ingredients for the sweet cucumber dressing. Cover with plastic wrap and leave to stand at room temperature while cooking the salmon.

Heat a barbecue hotplate or large heavy-based frying pan to medium–high, then lightly brush with oil. Cook the salmon cutlets for 2 minutes on each side, or until cooked but still a little pink in the middle — take care not to overcook the fish or it will be dry.

Transfer the salmon to warmed serving plates. Spoon the sweet cucumber dressing over the top and sprinkle with toasted nori strips. Serve with steamed rice.

note Pickled ginger and nori are Japanese ingredients available from Asian food stores and larger supermarkets.

Cucumbers contain silica, an essential component of healthy connective tissue and therefore very good for the skin. Over the centuries cucumber has been used topically to treat sunburn and eye-swelling, and it is also helpful in preventing fluid retention.

Cucumber, feta, mint and dill salad

SERVES 4

125 g ($4^1/2$ oz) feta cheese
4 Lebanese (short) cucumbers
1 small red onion, thinly sliced
$1^1/2$ tablespoons finely chopped dill
1 tablespoon dried mint
3 tablespoons olive oil
$1^1/2$ tablespoons lemon juice
crusty bread, to serve

Crumble the feta into 1 cm ($1/2$ inch) chunks and place in a large bowl. Cut the cucumbers into 1 cm ($1/2$ inch) lengths, then add to the bowl along with the onion and dill.

Grind the mint in a mortar and pestle, or force it through a sieve until powdered. Tip into a small bowl, add the olive oil and lemon juice and whisk until combined. Season with a little sea salt and freshly ground black pepper, pour over the salad and toss well. Serve with crusty bread.

Buying and storing

- Most varieties of lettuce are available from markets year-round.

- Choose lettuce with crisp-looking, brightly coloured heads and avoid any that are wilted.

- When choosing lettuce for a salad, most types are generally interchangeable, but when adding other ingredients, pick a leaf type that will suit them — floppy leaves won't go well with heavy ingredients such as potatoes, and crisp, firm leaves need a fairly robust dressing.

- Store in perforated plastic bags in the crisper of the fridge for up to 2 days.

lettuce

Undeniably one of the world's favourite vegetables, lettuce is at its best when served as the star ingredient of a simple salad. Whether providing the crunchy base to an anchovy-spiked caesar salad, a spicy Thai beef salad or a simple vinaigrette-dressed garden salad, a summer meal is unthinkable without a refreshing lettuce salad. Although it is usually eaten raw, lettuce can also be cooked. In France it is cooked with onions and peas in stock, in China, lettuce is a common cooked vegetable, while the Greeks use cos lettuce in a summery lemon and dill-infused lamb stew.

Lettuce teams well with ingredients such as avocado, tomatoes, cucumber, bread, cheese, herbs, mayonnaise, mustard, noodles, olive oil, chicken, beef, bacon, pork, prawns (shrimp), lobster and vinaigrette.

Varieties

There are hundreds of varieties of lettuce but for ease they can be divided into four groups.

Crisphead These are usually round and consist of tightly packed layers of juicy, crisp leaves. These include the iceberg (before widespread refrigeration, the lettuces were packed in boxes and covered with ice for transport and said to resemble icebergs), the British webb lettuce, and in the United States, the imperial, western and vanguard lettuces are popular.

Butterhead These have looser heads than crispheads, their leaves are soft and thin, and their flavour is mild and buttery. With soft, pale green leaves, butter lettuce is the definitive example of this variety, as well as the curly, reddish-tinged dark green-leaved mignonette. In the United States, the bibbs and boston varieties are popular.

Cos Also known as romaine, this lettuce originated on the Greek island of Cos, where it was discovered by the ancient Romans. It has narrow upright green leaves and a sweet, crisp flavour. Very young versions, sold as 'baby cos' or 'little gems', are also available.

Loose-leaf Distinguished by the fact that they do not form a distinct heart or head, these lettuces have soft leaves emanating directly from their core. The main varieties include the pale green- to rose-tinted oakleaf (feuille de chêne); the tender, frilly, light green or reddish-leaved coral; and the magenta-tipped and slightly bitter lollo rosso.

Preparation

Remove and discard the tough outer leaves before use and pull away the inner leaves. These need to be washed well before using — don't soak lettuce as this can make the leaves limp and soggy.

Dry the leaves well after washing using either paper towels or a salad spinner, as the salad dressing will adhere better if the leaves are dry. Always dress lettuce just before you serve it.

Caesar salad

SERVES 4–6

3 eggs
3 garlic cloves, crushed
2–3 anchovy fillets, drained
1 teaspoon worcestershire sauce
2 tablespoons lemon juice
1 teaspoon dijon mustard
185 ml (6 fl oz/^{3}/4 cup) olive oil
3 slices white bread
20 g (^{3}/4 oz) butter
1 tablespoon olive oil
3 bacon slices
1 large or 4 baby cos (romaine) lettuces,
   leaves separated, washed and dried
75 g (2^1/2 oz/^{3}/4 cup) shaved parmesan
   cheese

To make the dressing, put the eggs, garlic, anchovies, worcestershire sauce, lemon juice and mustard in a food processor and blend until smooth. With the motor running, add the olive oil in a thin, steady stream and process until the mixture is creamy. Season well, then cover the surface with plastic wrap to prevent a skin forming and set aside.

Remove the crusts from the bread and cut the bread into 1.5 cm (^{5}/8 inch) cubes. Heat the butter and olive oil in a frying pan, then add the bread and cook over medium heat for 5–8 minutes, or until golden all over, stirring often. Remove the croutons from the pan and drain well on paper towels.

Cook the bacon in the same frying pan for 3 minutes, or until golden and crisp, turning once. Break the bacon into bite-sized pieces.

Place the lettuce in a bowl, add the dressing and toss to coat. Mix in the croutons and bacon, scatter the parmesan over the salad and serve.

Green salad with lemon vinaigrette

SERVES 6

1 baby cos (romaine) lettuce
1 small butter lettuce
50 g (1^3/4 oz) watercress sprigs
100 g (3^1/2 oz) rocket (arugula) leaves

lemon vinaigrette
1 tablespoon finely chopped French shallot
2 teaspoons dijon mustard
^{1}/2 teaspoon sugar
1 tablespoon finely chopped basil
3 teaspoons lemon juice
1 tablespoon white wine vinegar
1 teaspoon lemon oil (see note)
4 tablespoons extra virgin olive oil

Discard the outer leaves from the lettuces, then separate the inner leaves. Wash the lettuce, watercress and rocket, then drain and dry thoroughly. Place in the fridge to chill.

To make the lemon vinaigrette, combine the shallot, mustard, sugar, basil, lemon juice and vinegar in a bowl and whisk together well. Mix the lemon oil and olive oil in a small jug and slowly add to the bowl in a thin stream, whisking constantly until a smooth, creamy dressing forms. Season to taste with sea salt and freshly ground black pepper.

Put the salad greens in a large bowl. Just before serving, drizzle the vinaigrette over the salad and toss gently to coat.

note Lemon oil is available from gourmet food stores, or make your own by steeping some lemon zest in some extra virgin olive oil. The flavour will become more intense the longer it steeps. Alternatively, add 1 teaspoon finely grated lemon zest to the vinaigrette and omit the lemon oil.

radish

Buying and storing

- Although radishes are available year-round, their peak season is from spring to autumn.
- Buy smooth, firm radishes that are not too large as they may be tough. Black radishes should feel very hard; any softness at all can indicate a 'woolly' interior.
- Store without leaves, as these will accelerate moisture loss, in a ventilated plastic bag in the fridge for up to 1 week.

Preparation

Generally, it is the crisp, tender 'spring' varieties of radishes that we mostly see, so look for these if you want to eat them raw in a salad. These need only trimming and washing. 'Winter' varieties are harder skinned and require peeling, and tend to be cooked rather than eaten raw.

A peppery root vegetable related to the mustard plant, radishes come in many varieties but are grouped under red, black or white. The hot spicy flavour of radishes comes from enzymes in the skin — peeling them reduces the effect, but that distinctive mustardy freshness is a large part of the radish's attraction. Red radishes are the mildest of the group. They are crisp and juicy, and usually eaten raw in salads. Black radishes, with their ivory flesh and dryish texture, have a stronger flavour and are often peeled before use to reduce their pungency; these can also be roasted. Large white radishes, called daikon, which means 'great root', are popular in Japanese cooking.

Salata baladi

SERVES 4 AS A SIDE SALAD

1$\frac{1}{2}$ tablespoons extra virgin olive oil
1$\frac{1}{2}$ tablespoons lemon juice
1 baby cos (romaine) lettuce, leaves torn
2 tomatoes, each cut into 8 wedges
1 small green capsicum (pepper), chopped
2 Lebanese (short) cucumbers, seeded and
   chopped
4 radishes, sliced
1 small red onion, thinly sliced
2 tablespoons chopped flat-leaf (Italian)
   parsley
1 small handful mint

To make the dressing, whisk together the
olive oil and lemon juice. Season well with
sea salt and freshly ground black pepper.

Put the lettuce, tomato, capsicum, cucumber,
radish, onion and herbs in a large serving
bowl and toss together well. Add the dressing
and toss again. Serve at once, while the salad
is crisp.

Zucchini, radish and feta salad

SERVES 4

1 tablespoon white wine vinegar
2 tablespoons olive oil
2–3 teaspoons wholegrain mustard
5 small zucchini (courgettes)
6 radishes
$\frac{1}{2}$ small red onion
1 small cos (romaine) lettuce
100 g (3$\frac{1}{2}$ oz/$\frac{2}{3}$ cup) crumbled feta cheese

To make the dressing, combine the vinegar,
olive oil and mustard. Season with sea salt
and freshly ground black pepper.

Using a vegetable peeler, shave the zucchini
into long strips. Place in a colander, sprinkle
with 2 teaspoons sea salt and set aside to
drain for 30 minutes. Do not rinse. Gently dry
with paper towels and put in a large bowl.

Finely slice the radishes and onion and add to
the bowl. Tear the inner leaves of the lettuce
into smaller pieces and add to the bowl with
the dressing and toss lightly. Transfer to a
serving dish and scatter the crumbled feta
cheese over the top before serving.

Daikon radishes are an Asian staple, particularly in Japan.
They have a firm, crisp flesh and a mild flavour, similar to that
of white turnip. Daikon can be diced and added to salads, or
used like a potato or turnip and added to soups, stews or stir-
fries. In Japan, grated raw daikon is formed into a small pile
and served with sashimi and tempura.

rocket

Buying and storing

- Rocket is available year-round, but is best during spring and early summer.

- Choose rocket with dark green, dry leaves that are not wilted or torn. Younger or smaller leaves are milder than the larger leaves, which can get quite hot and peppery.

- Rocket is sold in bunches and is best used on the day of purchase, as it wilts quickly. However, it will keep in the crisper of the fridge, wrapped loosely in plastic, for 2 days.

Preparation

Wash rocket thoroughly to remove any grit before using. Dry well using a salad spinner or by blotting the leaves on paper towels.

Called *rucola* in Italy, *roquette* in France and arugula in America, rocket is a slightly bitter salad leaf with a nutty, peppery flavour. Cultivated rocket has much larger, broader leaves (when young, it is sold as 'baby rocket'), while wild rocket has small, thin, serrated leaves. Rocket is used mainly as a salad ingredient, most notably in the Italian dish that simply combines rocket and shaved parmesan cheese, dressed with olive oil and seasoned with sea salt and black pepper. Rocket can be used as a pizza topping, added at the last minute to a creamy potato soup, or wilted in pasta sauces.

The peppery flavour of **rocket** pairs well with ingredients such as avocado, beef, lamb, seafood, pasta, balsamic vinegar, olive oil, lemon juice, cheese, garlic, pears, pine nuts, walnuts, radicchio and tomatoes.

Rocket tarts

MAKES 4

2 sheets frozen butter puff pastry, thawed
1 tablespoon olive oil
1/2 small onion, finely diced
1 large handful baby rocket (arugula)
3 eggs, lightly beaten
125 g (4 1/2 oz/1/2 cup) ricotta cheese
a pinch of ground nutmeg

Preheat the oven to 180°C (350°F/Gas 4). Grease four 10 cm (4 inch) loose-based tartlet tins.

Cut out four 15 cm (6 inch) circles from the puff pastry and use them to line tins. Prick the bases with a fork. Line the pastry shells with baking paper and half-fill with baking beads or rice. Bake the pastry for 15 minutes, then remove the paper and baking beads and bake for a further 5 minutes, or until the pastry is light golden. Remove from the oven and set aside.

Heat the olive oil in a frying pan over medium heat. Add the onion and sauté for 5 minutes, or until softened. Add the rocket and remove from the heat.

In a small bowl, combine the eggs, ricotta and nutmeg. Season with sea salt and freshly ground black pepper and stir to just combine; the ricotta will not be completely smooth. Stir in the rocket mixture.

Spoon the filling into the pastry shells and bake for 25 minutes, or until set. Serve warm or at room temperature.

Spaghetti with rocket and chilli

SERVES 4

500 g (1 lb 2 oz) spaghetti or spaghettini
2 tablespoons olive oil
2 teaspoons finely chopped small red chilli
450 g (1 lb) rocket (arugula)
1 tablespoon lemon juice

Bring a large saucepan of salted water to the boil. Cook the pasta until al dente, then drain and return it to the pan.

Meanwhile, heat the olive oil in a large heavy-based frying pan. Add the chilli and cook for 1 minute over low heat. Add the rocket and cook, stirring often, for a further 2–3 minutes, or until softened. Add the lemon juice and season with sea salt and freshly ground black pepper. Add the rocket mixture to the pasta and toss until combined. Serve immediately.

tomato

Although botanically classed as a fruit, the tomato — another great gift to the world's cuisines, especially Italian — is used in cooking as a vegetable. So thoroughly has the tomato been assimilated that it's difficult to image life in the Western world before it arrived in Naples in the sixteenth century. For many centuries it was believed the tomato was poisonous (not an unreasonable assumption given that it belongs to the nightshade family and its leaves are, in fact, toxic). Today there are over a thousand varieties grown worldwide, with America and Italy being the largest producers.

Varieties

There are too many varieties of tomatoes to list here; instead, they are grouped under general types. Variables include skin thickness, flesh density, juiciness and the size and quantity of seeds, making some more suitable for certain uses over others.

Round or salad These rounded, squat tomatoes are the best known and most used of all the tomato types. They are a good all-purpose tomato; their shape lends itself to slicing, dicing or cutting into wedges for salads, for stuffing and baking, or to cutting in half or quarters and roasting, grilling (broiling), pan-frying, or in sauces, stews and soups.

Roma (plum) Also called 'egg' or Italian tomatoes, these have thickish skin, an elongated shape and a large interior 'core'. They are great for cooking as they hold their shape well and their flesh is fairly dense and firm. Romas have fewer seeds than other types of tomatoes and have less juicy (but no less flavoursome), meaty flesh. Use them in pasta sauces, casseroles and soups, or for oven- or sun-drying.

Cherry These have the greatest amount of sugar of all the tomato types, so are good for eating raw in salads or salsas, or halve them and use to top a homemade pizza, or briefly roast or pan-fry and serve with fish, chicken or meat.

Beefsteak These 'giant' tomatoes have a unique ribbed shape and firm, juicy flesh. Because of their size, they are good for slicing and using in hamburgers or sandwiches, and are also a good cooking tomato.

Green Some heirloom varieties are green when ripe, but generally tomatoes sold as 'green' are those at the end of the season, which will fail to fully ripen in the cooler weather. Firm, not as juicy as their ripe counterparts and piquant-tasting, green tomatoes are wonderful when made into relishes and chutneys, and their suitability for frying is legendary.

Golden or yellow Most red varieties listed on the previous page have a golden equivalent. These taste more or less identical to red tomatoes and can be used in the same ways.

Buying and storing

- Because tomatoes are available year-round, they have almost lost their association as a summer fruit. However, do try to buy them from your local market in summer — you'll be rewarded with the sweet-acid, fragrant, juicy hit of a 'real' tomato, not often found in supermarket versions.

- Select the best tomatoes not by their colour but by their aroma — smell the stem end and you should be able to inhale strong 'tomatoey' aromas.

- Choose tomatoes that feel heavy for their size, are plump-looking and shiny and that 'give' slightly to pressure when pressed. Tomatoes with any soft spots or skin damage will quickly turn mouldy and spoil.

- Often tomatoes destined for the supermarket have been picked while still green. These are then ripened to redness using ethylene gas and, even though they may look red and delicious, they won't have a fully developed flavour, aroma or texture.

- At home, take the tomatoes out of any packaging and store them at cool room temperature (not in the fridge), and use within a few days. Tomatoes are a subtropical fruit and do not like the cold; refrigeration also diminishes their flavour.

- If you buy tomatoes that are slightly underripe, leave them somewhere warm and in full sunlight and they will fully ripen in a few days.

- If you are lucky enough to have a glut of tomatoes, these can be frozen for future use. Simply freeze them whole; once they thaw the skins will slip easily off and you just need to cut away the stem end. Alternatively, they can be frozen peeled and chopped or puréed.

Preparation

Peeling Many recipes ask for tomatoes to be peeled, especially if they are to be cooked. During the cooking process the tomatoes eventually shed their skin, which can be quite unpleasant and chewy when you encounter them in your dish.

Remove the tomato stems and cut a small cross in the base of each one with a knife. Put in a bowl and cover with boiling water. Leave for about 20 seconds or so (the timing is variable depending on tomato ripeness and variety), then remove with a slotted spoon and transfer to a bowl of iced water. Starting from the cross, pull the skin away. If the skin doesn't come away easily, the tomatoes have not been blanched for long enough (be careful not to leave them in the boiling water for too long, as this will make the flesh mushy and soggy).

Seeding If the recipe asks for the seeds to be removed, cut the peeled tomato in half horizontally and squeeze out the seeds with your hands or scoop them out with a teaspoon. You can remove the seeds from an unpeeled tomato the same way. Cut or dice as directed.

Caponata

SERVES 6 AS A SIDE DISH

1 kg (2 lb 4 oz) tomatoes
3 tablespoons olive oil
2 onions, sliced
2 red capsicums (peppers), thinly sliced
4 garlic cloves, finely chopped
4 celery stalks, sliced
1 large eggplant (about 500 g/1 lb 2 oz), diced
2 tablespoons thyme
2 tablespoons caster (superfine) sugar
125 ml (4 fl oz/$1/2$ cup) red wine vinegar
125 g ($4 1/2$ oz/$1/2$ cup) pitted green olives, rinsed well and drained
2 tablespoons capers, rinsed and drained

Bring a saucepan of water to the boil. Using a small knife, score a cross in the base of each tomato. Place the tomatoes in the boiling water for about 20 seconds, remove using a slotted spoon, then plunge into a bowl of iced water. Drain the tomatoes and peel the skins away from the cross. Cut the tomatoes in half, scoop out the seeds with a teaspoon and roughly chop the flesh.

Heat the olive oil in a large heavy-based frying pan. Add the onion, capsicum, garlic, celery and eggplant. Cover and cook over low heat for 20–30 minutes, or until tender, stirring occasionally. Season to taste with sea salt and freshly ground black pepper.

Remove the lid, add the chopped tomato and thyme and simmer, uncovered, for a further 15 minutes. Stir in the sugar, vinegar, olives and capers and mix well. Taste and season again if needed.

Serve the caponata warm or at room temperature as an accompaniment to grilled meats or as part of an antipasti selection. Caponata is also delicious stirred through drained, cooked pasta and served topped with grated pecorino cheese as a light lunch or dinner.

Tomatoes contain lycopene, a substance that gives tomatoes their red colour. Lycopene is a powerful antioxidant and is believed to be an effective protector against the development of cancers and heart disease. Lycopene is particularly abundant near the skin of the tomato and is said to be more readily absorbed when cooked with oils such as olive oil. Tomatoes also contain vitamins A and C and, when raw, vitamin E, and minerals such as potassium, calcium and folic acid.

Tomato bread soup

SERVES 4

750 g (1 lb 10 oz) vine-ripened tomatoes
1 loaf (about 450 g/1 lb) day-old, good-
    quality rustic bread (see note)
1 tablespoon olive oil
3 garlic cloves, crushed
1 tablespoon tomato paste (concentrated
    purée)
1.25 litres (44 fl oz/5 cups) vegetable stock
    or water
4 tablespoons torn basil leaves
2–3 tablespoons extra virgin olive oil,
    plus extra to serve

Peel and seed the tomatoes (page 160), then
roughly chop the flesh.

Trim the crust from the bread, then tear the
bread into 3 cm (1$\frac{1}{4}$ inch) chunks.

Heat the olive oil in a large saucepan. Add
the garlic, tomato and tomato paste. Bring to
a simmer, stirring occasionally, and cook for
15 minutes, or until the mixture has reduced.

Add the stock and bring to the boil over high
heat, stirring, for 2–3 minutes. Reduce the
heat to medium, add the bread and cook,
stirring often, for 5 minutes, or until the bread
softens and absorbs most of the liquid. Add
more stock if the soup is too thick.

Remove from the heat, stir in the basil and
extra virgin olive oil, and leave to stand for
5 minutes for the flavours to develop. Serve
drizzled with a little extra olive oil.

note It is important to use a quality loaf from
a good baker, as mass-produced breads do
not go slightly stale after 1 day. Their texture
is generally too soft and smooth to work well
in this recipe.

Fattoush salad

SERVES 4

1 large pitta bread, split
2 baby cos (romaine) lettuces, torn into
    bite-sized pieces
2 tomatoes, chopped
2 small Lebanese (short) cucumbers,
    chopped
4 spring onions (scallions), chopped
1 green capsicum (pepper), cut into large
    dice
1 large handful mint, roughly chopped
1 large handful coriander (cilantro),
    roughly chopped

dressing
3 tablespoons lemon juice
3 tablespoons olive oil
1 tablespoon sumac

Preheat the oven to 180°C (350°F/Gas 4).
Put the pitta bread on a baking tray and bake
for 5 minutes, or until golden and crisp.
Remove from the oven and cool. Break into
2 cm ($\frac{3}{4}$ inch) pieces.

To make the dressing, mix the lemon juice, oil
and sumac together and season with sea salt
and freshly ground black pepper.

In a serving bowl, toss the lettuce, tomato,
cucumber, spring onion, capsicum, mint and
coriander together. Crumble over the toasted
pitta bread, drizzle with the dressing and
serve immediately.

Slow-roasted balsamic tomatoes

MAKES 40 PIECES

10 firm roma (plum) tomatoes
8 garlic cloves, crushed
4 tablespoons caster (superfine) sugar
4 tablespoons torn basil
1 tablespoon chopped oregano
a few drops of good-quality balsamic vinegar

Preheat the oven to 140°C (275°F/Gas 1). Line two baking trays with baking paper. Cut each tomato lengthways into quarters and arrange in rows on the baking trays.

In a small bowl, mix together the garlic, sugar, basil, oregano and vinegar. Using your fingers, press a little of the mixture onto the sides of each tomato quarter and season with sea salt and freshly ground black pepper.

Bake for 2$1/2$ hours, or until the tomatoes are slightly shrivelled around the edges and semi-dried, but still soft in the middle. Serve warm or at room temperature as part of an antipasti selection, or with barbecued meats.

Store in an airtight container in the fridge for up to 1 week.

The Italian word for **tomato**, pomodoro, comes from the first name they coined for it, *pomi d'oro* (golden apple), hinting at its original colour, while the English name comes from the Aztec, *tomatl*.

Tabbouleh

SERVES 6

130 g (4$1/2$ oz/$3/4$ cup) burghul (bulgur)
3 tomatoes
1 telegraph (long) cucumber
4 spring onions (scallions), sliced
120 g (4$1/4$ oz/4 cups) chopped flat-leaf
   (Italian) parsley
2 large handfuls mint, chopped

dressing
4 tablespoons lemon juice
3 tablespoons olive oil
1 tablespoon extra virgin olive oil

Put the burghul in a bowl, cover with 500 ml (17 fl oz/2 cups) water. Set aside to soak for 1$1/2$ hours.

Cut the tomatoes in half, scoop out the seeds with a teaspoon and cut into 1 cm ($1/2$ inch) cubes. Cut the cucumber in half lengthways, remove the seeds with a teaspoon and cut the flesh into 1 cm ($1/2$ inch) cubes.

To make the dressing, place the lemon juice and 1$1/2$ teaspoons salt in a bowl and whisk until well combined. Season well with freshly ground black pepper and slowly whisk in the olive oil and extra virgin olive oil.

Drain the burghul and use your hands to squeeze out any excess water. Spread the burghul out on paper towels and leave to dry for 30 minutes. Put the dry burghul in a large salad bowl, add the tomato, cucumber, spring onion, parsley and mint, and toss well to combine. Before serving, pour the dressing over the salad and toss well.

Buying and storing

- Watercress is available year-round, but its peak season is spring.

- Choose glossy green watercress with no wilted or yellowing leaves. Darker leaves and thicker stems indicate it is older, and will probably have quite an intense flavour.

- Store in a plastic bag in the crisper of the fridge for up to 2 days, or place the stalks in a bowl or glass of water, cover the leaves loosely with a plastic bag and refrigerate.

Preparation

Pick through a bunch of watercress, removing the sprigs and the single, tender stalks that occur further down the central stems. Discard the thick, tough stems. Gently wash the picked watercress and pat dry.

watercress

Watercress, as its name would suggest, is an aquatic plant that is both cultivated and found growing in the wild in countryside streams and brooks. Its green leaves are characterised by a tangy mustard-like flavour, which becomes more pronounced as it matures. Watercress is often used as a garnish or tossed into green salads to add texture and flavour, it can be puréed along with potato to make a soup, used like wilted spinach in a quiche filling (cooking destroys its potency), and the English are fond of their watercress sandwiches, which may also include boiled eggs, chopped prawns or smoked salmon.

Watercress salad

SERVES 4–6

500 g (1 lb 2 oz) watercress
3 celery stalks
1 Lebanese (short) cucumber
3 oranges
1 red onion, thinly sliced and separated
    into rings
35 g (1 1/4 oz/3/4 cup) snipped chives
60 g (2 1/4 oz/1/2 cup) chopped pecans
    or walnuts

dressing
3 tablespoons olive oil
3 tablespoons lemon juice
2 teaspoons finely grated orange zest
1 teaspoon seeded mustard
1 tablespoon honey

To make the salad, wash and drain the
watercress, then break the watercress into
small sprigs, discarding the coarser stems.
Cut the celery into thin 5 cm (2 inch) long
sticks. Peel, halve and seed the cucumber
and cut into thin slices. Peel the oranges,
remove all the white pith and cut the oranges
into segments between the membrane.
Refrigerate until needed.

To make the dressing, combine the olive oil,
lemon juice, lemon zest, mustard, honey and
some freshly ground black pepper in a jar
with a lid. Shake vigorously to combine well.

Combine all the salad ingredients, except the
nuts, in a serving bowl. Pour the dressing over
and toss to combine. Sprinkle the salad with
the pecans before serving.

Beetroot and blue cheese salad

SERVES 4

1 tablespoon olive oil
50 g (1 3/4 oz/1/2 cup) pecans
1.3 kg (3 lb) small beetroot (beets), washed,
    trimmed and halved
250 g (9 oz) baby green beans, trimmed
120 g (4 1/4 oz) watercress, trimmed
2 tablespoons walnut oil
1 teaspoon honey
2 teaspoons finely grated orange zest
1 tablespoon cider vinegar
50 g (1 3/4 oz) firm blue cheese, such as
    stilton, crumbled

Heat the oil in a frying pan over medium–high
heat, add the pecans and cook, stirring often,
for 3 minutes, or until lightly toasted. Sprinkle
with salt and pepper. Remove from the heat
and place on paper towels to drain.

Line a steamer with baking paper, punch
holes in the paper, place the beetroot in the
steamer and cover with a lid. Set the steamer
over a saucepan of boiling water and steam
the beetroot for 30–35 minutes, or until
tender when pierced with a knife. Remove
from the steamer and cool. Remove the
baking paper from the steamer, then add the
beans to the steamer, cover and cook for
5–7 minutes, or until just tender. Remove the
beans and refresh under cold water.

Peel the beetroot, trim off any excess stem
and cut into wedges. Combine the pecans,
beans and watercress in a large bowl. Whisk
together the walnut oil, honey, orange zest
and vinegar in a bowl, then pour over the
salad. Add the beetroot and stir gently to just
combine. Season, then transfer to a serving
platter and sprinkle with the blue cheese.

pods & beans

bean

Green beans are native to tropical America and were cultivated in Mexico and Peru over 7000 years ago. Green beans are so-called because they are picked when young and still tender, and not for their colour, which can be light to deep green, yellow, russet or purple.

Varieties

Green These beans are variously known as French beans, 'snap' beans (it should 'snap' when you bend it), and 'string' beans because of the fibrous strings that run down the length of some varieties (this has mostly been bred out). Green beans are picked when immature — the 'green' refers to their immaturity when picked and not their colour. They also vary greatly in pod size and shape, from the tiny haricots verts (green string bean), to the large flat beans (sometimes called Italian beans), which can be up to 25 cm (10 inches) in length.

Green beans are best boiled in plenty of salted water or steamed; they can also be roasted in olive oil, which concentrates their sweet flavours. Flat beans are long, full flavoured and

meaty textured. They are suited to long cooking in soups, stews and braises and can even be barbecued until tender.

Wax Also called yellow or butter beans, these are actually just a pale variety of the green bean. Wax beans have a gentler flavour than green beans so serve them simply — steamed and dressed with butter.

Runner These flat beans should snap crisply when fresh, and most need to be stringed unless very young. The beans are often sliced before cooking, although this will cause most of the nutrients to leach out during cooking.

Broad (fava) These beans are enclosed in a soft downy substance, inside a large green pod. The first broad beans of spring are small and tender (these don't need peeling), but as the season progresses, they become larger and take on starchier, less sweet characteristics.

Young broad beans can be treated simply — steam them, drizzle with olive oil and serve with flakes of sea salt as a simple starter, or toss them through pasta with a saffron cream sauce. Older broad beans have a mealier texture and a less sweet flavour — use these for a purée with garlic and olive oil to spread on crostini, or cook them in a mixed spring vegetable soup or stew.

Snake (yard-long) These long beans are like green beans and are generally sold in bundles. Make sure any swellings in the pods are small — this will mean the beans are young and fresh.

Borlotti (cranberry) These beans have distinguishable cream and red pods with beans the same colour. It is the seeds that are eaten fresh, but these are often left to reach full maturity, at which point they are dried.

Simmer borlotti for 15–30 minutes, or until tender. Never boil them, as this will make them tougher. Borlotti beans are popular in Italy, where they are mainly used in soups or stewed with olive oil and garlic as a side dish.

Soya The most nutritious of all beans, these have been cultivated in their native China for thousands of years. The beans grow in green pods, which are covered with a slight fuzz. Inside the pods are pea-like seeds; these have a sweet, delicate flavour and buttery texture.

Preparation

Green, wax, runner and snake beans Check if your beans need stringing. Trim the 'top' end of each bean (where it was attached to the plant). It is not necessary to trim the 'tail' end.

To prepare beans for use in salads, it is best to slightly undercook them, drain well, then leave to cool. Some recipes suggest to plunge the cooked beans into iced water to cool — this preserves their green colour but compromises their flavour. Beans can be boiled in plenty of salted water for 4–7 minutes, depending on whether they are whole, halved or sliced. To roast whole, toss in olive oil and roast in a 180°C (350°F/Gas 4) oven for about 20 minutes.

Snake beans are chopped into smaller lengths for cooking and can be cooked as for green beans, although they are often stir-fried in curries and other Asian dishes.

Broad beans (fava) Very young broad beans can be eaten in their pods like snow peas (mangetout), but as they get older, the pods become tougher and the beans inside develop a grey, leathery skin. Generally, broad beans need to be removed from their pods before cooking, then peeled to remove the grey skins.

If you want to serve your beans as a simple side dish or in dips or purées where the beans need to be fully tender, simmer the beans until tender (about 10 minutes for young broad beans and 20 minutes for more mature beans). Drain and cool, then peel. Reheat briefly if necessary in a little simmering water or toss in a hot frying pan with some butter or olive oil before serving.

If you want to use broad beans in a dish such as risotto or soup where they will receive additional cooking, just blanch them for 2–3 minutes in boiling water. Drain and cool in a bowl of iced water. Drain again and peel off the skins.

Soya beans These need to be podded and this is easier to do after cooking. Boil them in their pods for 5 minutes, or until the beans inside are tender — taste one to check. Drain and cool, then remove from their pods and use in stir-fries, soups or salads, or refresh in iced water and serve in their pods as a snack (page 178).

Buying and storing

- Most varieties of beans are available year-round, however the peak season for green beans is from spring to summer; soya beans, runner and snake beans are best in summer; and buy broad beans from winter to spring.

- You should not be able to bend green beans; they should break crisply. The pods should be smooth, with no wrinkled areas, and uniformly coloured with almost imperceptible seed bumps underneath; large swellings where the seeds are indicate they are older.

- Choose broad beans with smooth, pale green pods. Avoid any that are bulgy: this indicates large, hard beans within. If possible, break a pod open and check that the white lining is moist and the beans have taut, smooth skins. When buying broad beans, note that 1 kg (2 lb 4 oz) of broad beans will yield about 300 g (10$\frac{1}{2}$ oz/1$\frac{1}{2}$ cups) of podded beans.

- Fresh soya beans are not widely available, although they are sold frozen from Asian food stores. Fresh soya beans should have bright green pods with healthy looking whitish or brown furry down.

- Store beans in a ventilated plastic bag in the fridge for 2–3 days.

Niçoise salad with green beans and seared tuna

SERVES 4

4 tablespoons olive oil
2 garlic cloves, crushed
1 teaspoon dijon mustard
1 tablespoon white wine vinegar
2 x 150 g (5 1/2 oz) tuna fillets
olive oil, for brushing
225 g (8 oz) green beans, trimmed
450 g (1 lb) new potatoes, scrubbed, halved
2 large handfuls torn butter lettuce leaves
8 cherry tomatoes, halved
12 kalamata olives
2 hard-boiled eggs
4 anchovies, halved
1 tablespoon capers, rinsed and drained

In a small bowl, whisk together the oil, garlic, mustard and vinegar. Season with sea salt and freshly ground black pepper; set aside.

Preheat a barbecue grill or a large heavy-based frying pan to medium–high. Brush the tuna with oil and cook for 1 1/2–2 minutes on each side, or until well browned but still a little rare in the middle. Transfer to a plate.

Cook the beans in a saucepan of boiling salted water for 3 minutes, or until just tender. Remove with tongs and drain. Add the potatoes and boil for 10 minutes, or until tender. Drain, cool slightly, then place in a shallow salad bowl with the lettuce leaves. Add 2 tablespoons of the dressing and toss to coat. Scatter the tomatoes over the top.

Slice the tuna on the diagonal into 5 mm (1/4 inch) slices; arrange over the tomatoes. Add the beans and olives. Shell the eggs, cut into wedges and add to the salad. Scatter the anchovies and capers over the top, drizzle with the remaining dressing and serve.

Green beans with tomato and olive oil

SERVES 4 AS A SIDE DISH

4 tablespoons olive oil
1 large onion, chopped
3 garlic cloves, finely chopped
400 g (14 oz) tin chopped tomatoes
1/2 teaspoon sugar
750 g (1 lb 10 oz) green beans, trimmed
3 tablespoons chopped flat-leaf (Italian)
    parsley

Heat the olive oil in a large frying pan. Add the onion and sauté over medium heat for 5 minutes, or until softened. Add the garlic and cook for a further 30 seconds. Add the tomatoes, sugar and 125 ml (4 fl oz/1/2 cup) water, then season with sea salt and freshly ground black pepper.

Bring to the boil, then reduce the heat and simmer for 10 minutes, or until the liquid has reduced slightly.

Add the beans, then partially cover and simmer for a further 10 minutes, or until the beans are tender and the tomato mixture is pulpy. Stir in the parsley, check the seasoning and serve.

Borlotti bean moussaka

SERVES 6

250 g (9 oz/1 1/4 cups) dried borlotti beans
2 large eggplants (aubergines)
4 tablespoons olive oil
1 onion, chopped
1 garlic clove, crushed
125 g (4 1/2 oz) button mushrooms, wiped
    clean and sliced
250 ml (9 fl oz/1 cup) red wine
2 x 440 g (15 1/2 oz) tins peeled diced
    tomatoes
1 tablespoon tomato paste (concentrated
    purée)
1 tablespoon chopped oregano

topping
250 g (9 oz/1 cup) plain yoghurt
4 eggs, lightly beaten
500 ml (17 fl oz/2 cups) milk
1/4 teaspoon sweet paprika
50 g (1 3/4 oz/1/2 cup) grated parmesan
    cheese
40 g (1 1/2 oz/1/2 cup) fresh breadcrumbs

Soak the borlotti beans in cold water
overnight. Drain well and rinse. Transfer the
beans to a saucepan, cover with cold water
and bring to the boil. Reduce the heat to a
simmer and cook the beans over low heat
for 1 1/2 hours, or until tender. Drain well.

Slice the eggplant, sprinkle with salt and
allow to stand for 30 minutes.

Preheat the grill (broiler) to medium–high.
Rinse the eggplant, then pat dry on paper
towels. Brush the eggplant all over with some
of the olive oil, then grill (broil) for 3 minutes
on each side, or until golden. Drain the
eggplant on paper towels.

Heat the remaining oil in a large heavy-based
saucepan. Add the onion and garlic and cook
over medium heat for 4–5 minutes, or until
the onion is golden. Add the mushrooms and
cook for 3 minutes, or until lightly browned.
Add the wine, increase the heat to high and
cook for 2–3 minutes. Stir in the tomatoes,
tomato paste and oregano. Bring the mixture
to the boil, then reduce the heat and simmer
for 40 minutes, or until the mixture has
reduced and thickened.

Preheat the oven to 200°C (400°F/Gas 6).

Spoon the borlotti beans into a large
ovenproof dish and top with the tomato
sauce and eggplant slices.

To make the topping, whisk together the
yoghurt, eggs, milk and paprika, then pour
over the mixture in the dish. Allow to stand
for 10 minutes.

Combine the parmesan and breadcrumbs
in a small bowl and then sprinkle over the
moussaka. Transfer to the oven and bake for
50–55 minutes, or until hot, bubbling and
golden on top.

Chicken, artichoke and broad bean stew

SERVES 4

Originally native to Europe, broad beans are now eaten worldwide, and are popular in the Middle East, China, South America and the Mediterranean. This hearty Mediterranean stew combines broad beans with artichokes and rosemary.

60 g (2¼ oz/½ cup) plain (all-purpose) flour
8 chicken thighs on the bone, skin on
2 tablespoons olive oil
1 large red onion, cut into small wedges
125 ml (4 fl oz/½ cup) dry white wine
250 ml (9 fl oz/1 cup) chicken stock
2 teaspoons finely chopped rosemary
340 g (12 oz) jar marinated artichoke hearts, drained well and cut into quarters
155 g (5½ oz) frozen broad (fava) beans, peeled

potato mash
800 g (1 lb 12 oz) potatoes, peeled and cut into large chunks
60 g (2¼ oz) butter
3 tablespoons chicken stock

Season the flour with sea salt and freshly ground black pepper. Dust the chicken thighs in the flour, shaking off the excess.

Heat the olive oil in a saucepan or flameproof casserole dish. Add the chicken in batches and brown over medium heat for 8 minutes, turning once. Remove the chicken and drain on paper towels.

Add the onion to the pan and sauté for 3–4 minutes, or until softened but not browned. Increase the heat to high, add the wine and boil for 2 minutes, or until reduced to a syrupy consistency. Stir in the stock and bring the mixture just to the boil.

Return the chicken to the pan and add the rosemary. Reduce the heat to low, then cover and simmer for 45 minutes.

Add the artichokes to the pan, increase the heat to high and return to the boil, then reduce the heat and simmer, uncovered, for 10–15 minutes. Add the broad beans and cook for a further 5 minutes.

Meanwhile, make the potato mash. Cook the potato in a saucepan of boiling salted water for 15–20 minutes, or until tender. Drain, then return the potato to the saucepan. Add the butter and stock and mash well using a potato masher.

Spoon the mashed potato into four warmed shallow bowls, then spoon the stew over the mash and serve.

The **soya bean** is the only vegetable to contain all nine essential amino acids, making it a source of complete protein and therefore an important ingredient in vegetarian diets. Soya beans also contain other nutrients, including antioxidants, phytochemicals, essential fatty acids, calcium and B vitamins.

Butter beans with sun-dried tomatoes and capers

SERVES 4 AS A SIDE DISH

2 sun-dried tomatoes in oil, plus 1 teaspoon of oil, extra
2 teaspoons capers, rinsed and drained
250 g (9 oz) young butter beans, trimmed
1 teaspoon light olive oil
finely grated zest of 1 lemon

Slice the sun-dried tomatoes into long, thin strips. Heat the extra oil from the sun-dried tomatoes in a small frying pan over medium heat and fry the capers, stirring often, for about 1 minute, or until darkened and crisp. Drain on paper towels.

Bring a saucepan of salted water to the boil. Add the beans and simmer for 3–4 minutes, or until just tender.

Drain, season with freshly ground black pepper and toss with the sun-dried tomatoes, capers, olive oil and lemon zest. Serve the beans hot or at room temperature.

Edamame

SERVES 4–6

500 g (1 lb 2 oz) fresh or frozen soya beans in pods (see note)
1–2 teaspoons instant dashi granules

Rub the fresh soya beans with salt between your hands to rub off the fine hairy fibres. Rinse the pods. If you are using frozen soya beans, this step is not necessary.

Combine the dashi granules with 1.25 litres (44 fl oz/5 cups) water in a saucepan and bring to the boil over high heat, stirring to dissolve the granules. Add the soya beans and cook for 6–8 minutes if using fresh or 3–4 minutes if using frozen, or until tender but still bright green. Drain well. You can serve the edamame either warm, at room temperature or chilled.

Sprinkle with lots of salt. To eat, simply suck the beans out of the pods and throw the pods away. Supply a bowl to collect the empty pods. Serve as a snack with a cold beer or glass of wine.

note Fresh soya beans in the pod, known as edamame in Japan, where the dish originated, are available fresh when in season or can be found frozen in your local Asian grocery store.

Soya bean, snow pea and prawn noodle stir-fry

SERVES 4

sauce
4 tablespoons red wine vinegar
4 tablespoons kecap manis
2 tablespoons soy sauce
2 tablespoons sesame oil
2 tablespoons grated ginger
1 tablespoon sweet chilli sauce
1 garlic clove, crushed

250 g (9 oz) egg noodles
150 g (5^1/$_2$ oz) raw prawns (shrimp)
1 tablespoon oil
200 g (7 oz/2 cups) small snow peas (mangetout)
1 red capsicum (pepper), seeded and thinly sliced
80 g (2^3/$_4$ oz/1^1/$_3$ cups) soya beans
6 spring onions (scallions), sliced
2 large handfuls coriander (cilantro)

To make the sauce, combine the vinegar, kecap manis, soy sauce, sesame oil, ginger, sweet chilli sauce and garlic and set aside.

Cook the egg noodles in boiling water for 2 minutes, then drain and set aside.

Peel and devein the prawns, leaving some with their tails on, if desired.

Heat the oil in a large wok over high heat, then add the prawns, snow peas, capsicum, soya beans and spring onion. Stir-fry for about 1 minute, or until the prawns are just opaque, then add the noodles and toss for 20–30 seconds. Pour in the sauce, then add the coriander, stir to combine and heat through, then serve.

Tamari roasted almonds with spicy green beans

SERVES 4–6

125 g (4^1/$_2$ oz/^{3}/$_4$ cup) almonds
tamari, for soaking
1 tablespoon oil
3 tablespoons sesame oil
500 g (1 lb 2 oz/2^1/$_2$ cups) rice
1 litre (35 fl oz/4 cups) boiling water
1 long red chilli, seeded and finely chopped
2 cm (^{3}/$_4$ inch) piece ginger, peeled and grated
2 garlic cloves, crushed
375 g (13 oz) green beans, cut into 5 cm (2 inch) lengths
125 ml (4 fl oz/^{1}/$_2$ cup) hoisin sauce
1 tablespoon soft brown sugar
2 tablespoons mirin

Soak the almonds in a bowl, with enough tamari to cover, for 30 minutes. Drain and dry with paper towels. Heat the oil in a non-stick frying pan over medium heat. Add the almonds and toss for 2–3 minutes, then remove and drain on paper towels. Roughly chop and set aside.

Preheat the oven to 200°C (400°F/Gas 6). Heat 1 tablespoon of sesame oil in a deep baking dish, add the rice and stir until well coated. Stir in the boiling water. Cover and transfer to the oven. Cook for 20 minutes, or until all the water has absorbed. Keep warm.

Meanwhile, heat the remaining sesame oil in a wok or large frying pan and cook the chilli, ginger and garlic for 1 minute. Add the beans, hoisin sauce and sugar and cook for a few more minutes. Stir in the mirin and cook for 1 minute, or until the beans are just tender. Remove the wok from the heat and stir in the almonds. Serve with the rice.

pea

Peas are the seeds of a legume and were once more valued in dried form than fresh. About 95 per cent of the world's pea crop is either frozen or tinned, so fresh peas really are a seasonal treat, but they must be bought at their optimum time — if harvested too late, they become dry and less sweet because their sugars convert into starch. In some cases, frozen peas — processed just after harvesting — are actually fresher than freshly podded peas. This is because fresh peas deteriorate so quickly. The most common pea is the garden pea, also called the English pea.

Types of garden peas

Snow peas (mangetout) The most common type of garden pea eaten in China and Japan today, snow peas are eaten pod and all (top and tail before eating). There are two types, those with a flat, thin pod (snow peas) and those with a more rounded pod (sugarsnap or snap peas).

Snow peas suit steaming, stir-frying and quick blanching — serve either as a simple side dish for meats or fish, or eat raw in salads. Use in Asian stir-fries teamed with classic Asian flavours such as ginger, garlic, chilli, fish sauce and soy sauce. Sugarsnap peas are more developed than snow peas and can be used whole, in stir-fries or noodle dishes.

Petits pois Not a different variety but these are peas that have been harvested young. The peas are shelled before cooking.

Buying and storing

- The best time to buy fresh peas is from spring to early summer.

- Ripe pea pods are bright green, shiny and plump and contain up to ten round seeds, or peas. Avoid ones that are dull-looking or yellowish, as these are old and will have lost moisture.

- Purchase pods that don't rattle when you give them a gentle shake — the peas inside should fill the pods quite tightly but should not cause the unopened pods to bulge (these peas will be too mature).

- Buy snow peas with bright, light green skin and avoid any really large ones, as these tend to be tough. If you see snow peas that still have a few petals at the stem end, choose these as this indicates they are very fresh.

- Sugarsnap peas should have taut, bright green pods and give out a distinct 'snap' when broken in two. Avoid any that are even a little flaccid or damp-looking.

- Peas, snow peas and sugarsnap peas deteriorate rapidly once picked, so use on the day of purchase if possible, or store for up to 2 days in the fridge, sealed in a plastic bag. Podded, blanched peas can be frozen for up to 2 months.

- When buying fresh peas in the pod, note that 1.25 kg (2 lb 12 oz) will yield sufficient peas for four people, as a side dish.

Preparation

To shell peas, split the pods open along their seam, then run your thumb under the peas to dislodge them into a bowl. They can then be simmered in a minimum of water for no more than 3 minutes — overcooked peas are not pleasant. Toss them with butter or olive oil before serving.

For snow peas, snip off the stem end and pull away the string that runs down the side of the pod, then cook for 1–2 minutes, either by steaming, stir-frying or simmering.

Some varieties of sugarsnap peas need to be stringed before cooking — just break off the stem end and pull off the strings. They can then be steamed for about 4 minutes or boiled for 2 minutes. Blanch them briefly before stir-frying or using in salads — although they can be served raw, cooking them briefly first brings out their flavour.

Peas make a flavoursome and colourful mash, either on their own or mashed with potatoes (add a little mint too, if you like). Otherwise, add peas to soups such as minestrone, pasta dishes, stir-fries, omelettes and frittatas, or very fresh young peas can be eaten raw in salads.

Steamed snow peas with blood orange mayonnaise

SERVES 4 AS A SIDE DISH

blood orange mayonnaise
1 egg yolk
1 teaspoon white wine vinegar
1/2 teaspoon dijon mustard
125 ml (4 fl oz/1/2 cup) olive oil
2 tablespoons blood orange juice

150 g (5 1/2 oz/1 1/2 cups) snow peas
   (mangetout)
2 cm (3/4 inch) piece ginger, peeled and
   cut into thin strips
1 orange
1/2 celery stalk, sliced
4 spring onions (scallions), sliced

To make the blood orange mayonnaise, whisk together the egg yolk, vinegar and mustard. Whisking constantly, gradually drizzle in the olive oil until you have a thick emulsion. Stir in the blood orange juice, to taste, then season with sea salt and freshly ground black pepper.

Trim the snow peas and put them in a steamer along with three or four thin strips of ginger. Cover and steam over a rolling boil for 2–5 minutes, or until tender.

Cut a thin slice off the top and bottom of the orange. Using a small knife, slice off the skin and bitter white pith. Holding the orange over a bowl to catch the juices, cut between the white membranes to remove the segments.

Transfer the snow peas to a bowl along with the celery, spring onions and orange segments. Serve with the mayonnaise.

Risi e bisi

SERVES 4

1.5 litres (52 fl oz/6 cups) chicken
   or vegetable stock
2 teaspoons olive oil
40 g (1 1/2 oz) butter
1 small onion, finely chopped
80 g (2 3/4 oz) pancetta, cut into small pieces
2 tablespoons chopped parsley
375 g (13 oz/2 1/2 cups) shelled young peas
220 g (7 3/4 oz/1 cup) risotto (arborio) rice
50 g (1 3/4 oz/1/2 cup) grated parmesan
   cheese

Put the stock in a saucepan, bring to the boil and then maintain at a low simmer.

Heat the olive oil and half the butter in a wide, large heavy-based saucepan and cook the onion and pancetta over low heat for 5 minutes, or until softened. Stir in the parsley and peas and add 2 ladlefuls of the stock. Simmer for 6–8 minutes.

Add the rice and the remaining stock to the pan. Simmer for 12–15 minutes, or until the rice is al dente and most of the stock has been absorbed. Stir in the remaining butter and parmesan. Season with sea salt and freshly ground black pepper, and serve.

sweet corn

No food is more emblematic of summer than corn. With its bright yellow kernels and sweet juice, sweet corn is used in all manner of ways. In America, where it is called maize, corn has reached almost legendary status, and recipes for it abound. The Native Americans believed corn to be a gift from their gods; their name for it is the source of the English word 'maize'.

Buying and storing

- Although sweet corn is available year-round, its peak season is from spring to summer.
- Choose corn cobs still in their husks, with a fresh-looking glossy silk and fat, shiny kernels. The cobs should feel and look moist and plump. Corn loses its sweetness and nutritional benefits quickly, so use the day you buy it, or store corn in its husks in the fridge for 2–3 days only.

Preparation

Before cooking, first strip away the husk and the silk, rinsing to remove any stray silk strands.

To remove the kernels for cooking, hold the cob upright and run a sharp knife down the cob — the kernels will come away neatly. One corn cob will yield about 100 g (3½ oz/½ cup) of kernels.

The kernels can then be cooked in boiling water for several minutes until tender, then tossed with butter. Corn kernels can be used in fritters, soups and risottos, or baked into cornbread. Simmer corn kernels in cream until the cream has reduced and thickened and spoon the creamed corn over grilled meats or chicken.

Boiling The classic way to enjoy corn is 'on the cob'. Simmer the cobs in water (don't add salt as it toughens the kernels) for 5 minutes until tender, drain well, then serve with butter, sea salt and pepper. Some cooks add a little milk or sugar to the water to retain flavour and softness.

Barbecuing Wrap the cobs in foil with some butter and cook on the barbecue for 20 minutes, turning it often. Alternatively, strip the husks off the cobs, remove the silk and then replace the husk. Soak it briefly in water and cook on the barbecue. The husk will burn off as it cooks.

Baking Wrap the corn cobs in foil and bake in a 180°C (350°F/Gas 4) oven for 25 minutes. Serve with garlic- or parsley-flavoured butter.

Corn chowder

SERVES 8

90 g (3¼ oz) butter
2 large onions, finely chopped
1 garlic clove, crushed
2 teaspoons cumin seeds
1 litre (35 fl oz/4 cups) vegetable stock
2 potatoes, peeled and chopped
250 g (9 oz/1 cup) tin creamed corn
400 g (14 oz/2 cups) sweet corn kernels
    (about 4 cobs)
3 tablespoons chopped parsley
125 g (4½ oz/1 cup) grated cheddar
    cheese
3 tablespoons cream (optional)
2 tablespoons snipped chives

Melt the butter in a large heavy-based saucepan. Add the onion and sauté over medium–high heat for 5 minutes, or until the onion is golden.

Add the garlic and cumin seeds to the pan and cook for 1 minute, stirring constantly, then pour in the stock and bring to the boil. Add the potato, then reduce the heat and simmer for 10 minutes.

Add the creamed corn, corn kernels and parsley. Bring to the boil, then reduce the heat and simmer for 10 minutes.

Stir in the cheese and season to taste with sea salt and freshly ground black pepper. Stir in the cream, if using, and heat gently until the cheese melts. Serve immediately, sprinkled with snipped chives.

Barbecued corn in the husk

SERVES 8

8 sweet corn cobs, in their husks
125 ml (4 fl oz/½ cup) olive oil,
    plus extra for brushing
6 garlic cloves, finely chopped
4 tablespoons chopped flat-leaf (Italian)
    parsley
butter, to serve

Peel back the corn husks, leaving them attached to the corn at the end. Pull off the corn silks, then wash the corn and pat dry with paper towels.

In a small bowl, mix together the olive oil, garlic, parsley and some sea salt and freshly ground black pepper. Brush each corn cob with some of the mixture, then pull the husks back up over the corn and tie together at the top with string.

Working in batches, put the corn in a steamer and steam over a saucepan of boiling water for 5 minutes, then remove using tongs and pat dry.

Meanwhile, heat a barbecue grill or plate to medium and lightly brush with oil. Barbecue the corn for 20 minutes, turning regularly and occasionally spraying with water to keep the corn moist. Serve the corn hot, with knobs of butter.

Corn and polenta pancakes with bacon and maple syrup

SERVES 4

90 g (3 1/4 oz/3/4 cup) self-raising flour
110 g (3 3/4 oz/3/4 cup) fine polenta
300 g (10 1/2 oz/1 1/2 cups) sweet corn
    kernels (about 3 cobs)
375 ml (13 fl oz/1 1/2 cups) milk
olive oil, for pan-frying
8 rindless bacon slices
175 g (6 oz/1/2 cup) maple syrup
    or golden syrup

Preheat the oven to 120°C (235°F/Gas 1/2).

Sift the flour into a bowl and stir in the polenta. Add the corn kernels and 250 ml (9 fl oz/1 cup) of the milk and stir until just combined. Season with salt and freshly ground black pepper, then stir in the remaining milk.

Heat 3 tablespoons olive oil in a large frying pan. Using half the batter for the first batch, spoon it into the pan to make four 9 cm (3 1/2 inch) pancakes. Cook over medium heat for 2 minutes on each side, or until golden and cooked through. Drain on paper towels and place in the oven to keep warm while cooking the remaining four pancakes, adding more oil if necessary. Transfer to the oven to keep warm.

Add the bacon to the same pan and cook for 5 minutes, or until browned and crisp.

Put two corn pancakes and two bacon slices on each plate and serve drizzled with a little maple syrup.

Creamy corn bake

SERVES 4

250 g (9 oz/1 cup) crème fraîche
1 egg
25 g (1 oz/1/4 cup) grated parmesan cheese
30 g (1 oz/1/4 cup) self-raising flour
300 g (10 1/2 oz/1 1/2 cups) sweet corn
    kernels (about 3 cobs)
a pinch of cayenne pepper
2 tablespoons grated parmesan cheese,
    extra
40 g (1 1/2 oz) butter

Preheat the oven to 190°C (375°F/Gas 5). Lightly grease an 18 cm (7 inch) square baking dish.

Combine the crème fraîche, egg, parmesan and flour in a large bowl. Add the corn kernels and cayenne pepper and season with sea salt and freshly ground black pepper. Stir to combine.

Spoon the corn mixture into the prepared dish. Sprinkle with the extra parmesan and dot with the butter. Bake for 30–35 minutes, or until the mixture is set and the top is golden brown. Serve hot, as a side dish or with a salad and bread.

Index

Published in 2009 by Murdoch Books Pty Limited

Murdoch Books Australia
Pier 8/9
23 Hickson Road
Millers Point NSW 2000
Phone: +61 (0) 2 8220 2000
Fax: +61 (0) 2 8220 2558
www.murdochbooks.com.au

Murdoch Books UK Limited
Erico House
6th Floor, 93–99 Upper Richmond Road
Putney, London SW15 2TG
Phone: +44 (0) 20 8785 5995
Fax: +44 (0) 20 8785 5985
www.murdochbooks.co.uk

Chief Executive: Juliet Rogers
Publishing Director: Kay Scarlett

Commissioning editor: Lynn Lewis
Senior designer: Heather Menzies
Design concept and design: Jacqueline Richards
Editor: Kim Rowney
Additional text: Leanne Kitchen
Production: Elizabeth Malcolm
Photographers: Alan Benson, Steve Brown, Natasha Milne, Stuart Scott, George Seper
Cover photography: Stuart Scott
Stylists: Marie-Hélène Clauzon, Sarah O'Brien
Food preparation: Joanne Glynn
Recipes: Murdoch Books test kitchen

NOTE: For fan-forced ovens, set the oven temperature to 20°C (35°F) lower than indicated in the recipe. We have used 20 ml tablespoon measures.
Seasonal availabilities are given only as a guide; regional differences may apply.

National Library of Australia Cataloguing-in-Publication entry
Title: Cooking from the market – vegetables
ISBN: 9781741965469 (pbk.)
Notes: Includes index.
Subjects: Cookery (Vegetables)
  Vegetables--Handling.
  Vegetables--History.
Other Authors/Contributors: Lewis, Lynn.
Dewey Number: 641.65
A catalogue record for this book is available from the British Library.

PRINTED IN CHINA